BRITISH GEOLOGICAL SURVEY
Scotland

A. DAVIES,
A. D. McADAM and
I. B. CAMERON

CONTRIBUTORS

R. W. Elliot, D. K. Graham,
I. Strachan and R. B. Wilson

Geology of the Dunbar district

Memoir for 1:50 000 Sheet 33E and part of Sheet 41

BRITISH GEOLOGICAL SURVEY
Natural Environment Research Council

LONDON: HER MAJESTY'S STATIONERY OFFICE   1986

iv

Bibliographical reference

DAVIES, A., McADAM, A. D., and CAMERON, I. B. 1986. Geology of the Dunbar district. *Mem. Br. Geol. Surv.*, Sheet 33E and part of Sheet 41 (Scotland), 69 pp.

Authors

A. DAVIES, BSc
A. D. McADAM, BSc, MIGeol
I. B. CAMERON BSc
British Geological Survey, Edinburgh

Contributors

R. W. Elliot, BSc, D. K. Graham, BA, and R. B. Wilson, DSc, FRSE
British Geological Survey, Edinburgh

I. Strachan, BSc, PhD
Formerly University of Birmingham

Other publications of the Survey dealing with this district and adjoining districts

BOOKS

British Regional Geology
The Midland Valley of Scotland, 1985
The South of Scotland, 1971

Memoirs
Edinburgh (32), 1962
Haddington (33W), 1985

MAPS

1:625 000
Solid geology (North sheet)
Quaternary geology (North sheet)
Aeromagnetic (North sheet)

1:50 000
Sheet 33W (Haddington) Solid 1983; Drift 1978
Sheet 33E (Dunbar)   Solid 1980; Drift 1978
Sheet 34   (Eyemouth) Solid 1982; Drift 1983

Printed for Her Majesty's Stationery Office by Linneys Colour Print, Mansfield

Dd 738612 C20 6/86 47984

Geology of the Dunbar district

The district described includes that part of the East Lothian coast about 10 km on either side of Dunbar and extends southwards across the north-eastern part of the Lammermuir Hills. The district is covered by Sheet 33E and a small part of Sheet 41E of the geological map of Scotland.

The introductory chapter gives an outline of the physical features of the district and of the geological history. The Lower Palaeozoic rocks of the Lammermuir Hills are then described with particular reference to their structure, stratigraphy and lithology. Chapters then follow dealing with the sediments of Lower Devonian, Devono-Carboniferous and Lower Carboniferous age. A section on the palaeontology of the Lower Carboniferous rocks is followed by details of the igneous rocks in the district which comprise the Garleton Hills Volcanic Rocks, of Lower Carboniferous age, and the in-trusive rocks.

A chapter is devoted to the geological structure of the district and this is followed by a discussion of the Quaternary deposits including their history of deposition and their present distribution. The final chapter deals with the economic geology and the various natural resources present.

Cliffs in volcanic agglomerate, Dunbar Harbour; ruins of Dunbar Castle (D 3657) *Frontispiece*

CONTENTS

FIGURES

PLATES

TABLE

PREFACE

The district described in this memoir is covered by Sheet 33E and part of Sheet 41 of the 1:50 000 Geological Map of Scotland. It forms part of the first area to be mapped by the Geological Survey in Scotland. It was surveyed in the late 1850's by A. C. Ramsay, H. H. Howell and A. Geikie and the first edition of the map (Sheet 33) appeared in 1860 followed by the explanatory memoir in 1866.

Between 1888 and 1896 B. N. Peach and J. Horne revised the Lower Palaeozoic rocks and from 1902 to 1904 the post-Silurian rocks were re-surveyed by E. B. Bailey, G. Barrow, C. T. Clough and H. B. Maufe. The second editions of the map and memoir were published in 1910.

A full revision of the district was carried out between 1961 and 1968 by Mr I. B. Cameron, Mr A. Davies, Dr M. F. Howells and Mr A. D. McAdam with Mr R. A. Eden as District Geologist. On the 1:63 360 scale the part of the district on Sheet 41 was published in 1970. The Drift edition of the whole district was published in 1978 and the Solid edition in 1980 on the 1:50 000 scale.

During the survey, cored boreholes were sunk at East Linton, Skateraw and Birnieknowes and the records obtained were very useful in elucidating the stratigraphy of the Carboniferous rocks. Since the publication of the previous edition of the memoir, numerous papers have been published on various topics concerning the district and these are referred to in the text. The fossils from the field localities were collected by Mr P. J. Brand and Mr D. K. Graham.

The present memoir was written mainly by Mr A. Davies, Mr A. D. McAdam and Mr I. B. Cameron. Mr R. W. Elliot contributed the sections on the petrography of the igneous rocks. Dr I. Strachan, formerly of the University of Birmingham, identified the graptolites and contributed a section on their stratigraphical significance. Dr R. B. Wilson wrote the chapter on Carboniferous palaeontology and Mr D. K. Graham wrote the section on Quaternary fossils. Details of the groundwater resources were contributed by Mr N. S. Robins. The photographs were taken by Mr T. S. Bain, Mr A. F. Christie and Mr F. I. MacTaggart and the plates depicting Carboniferous and Quaternary fossils were prepared by Messrs P. J. Brand and D. K. Graham respectively. The memoir was edited by Dr R. B. Wilson.

G. INNES LUMSDEN, FRSE
Director

British Geological Survey
Keyworth
Nottingham   NG12 5GG

10 December 1985

GEOLOGICAL SEQUENCE

The geological formations which occur within the district are summarised below:

SUPERFICIAL DEPOSITS (DRIFT)

Quaternary (Recent and Pleistocene)

Blown sand
Peat
Alluvium and lake deposits
Present beach and saltmarsh deposits
Post-Glacial marine and estuarine deposits
Late-Glacial marine and estuarine deposits
Glacial meltwater deposits
Till

SOLID FORMATIONS

		Generalised thickness m
Carboniferous		
DINANTIAN		
Lower Limestone Group	cyclic sequence of marine limestones, calcareous mudstones, siltstones, sandstones, seatclays and thin coals	70
Calciferous Sandstone Measures	upper part mainly sandstone with two marine limestones near the top and a few marine shell beds lower down; middle part of basaltic and trachytic lavas and tuffs of the Garleton Hills Volcanic Rocks in north-western part of district, possibly represented south-east of the Lammermuir Fault by isolated, thin developments; lower part mainly cementstones, mudstones and sandstones	400
Devono-Carboniferous		
Upper Old Red Sandstone	mainly red and red-brown sandstones, some siltstone and mudstone; cornstones characteristic of upper part; thin, lenticular basal conglomerate	400
Unconformity		
Lower Devonian		
Great Conglomerate	red-brown and purple-brown greywacke conglomerates, numerous fractured clasts, sandstone lenses	350
Unconformity		
Silurian		
Llandovery	greywackes, siltstones, shales and mudstones, a few graptolitic-shale bands	not known
Ordovician		
?Llandeilo-Caradoc	mainly greywackes and shales with graptolite-bearing shales, cherty shales, siltstones and conglomerates present	not known
Igneous intrusions		
Late Westphalian–Stephanian	quartz-dolerite and tholeiite dykes	
?Namurian	olivine-dolerite/teschenite suite	
?Dinantian to Stephanian	monchiquite/basanite suite	
Dinantian	felsite and phonolite sills, dykes and plugs, agglomerate and tuff	
Lower Devonian	quartz-porphyry, porphyrite, lamprophyre and plagiophyre dykes and other minor intrusions; granodiorite and norite major intrusions	

SIX-INCH MAPS

NOTES

The geological six-inch maps covering, wholly or in part, the 1:50 000 Dunbar (33E) Sheet are listed below with the names of the surveyors (I. B. Cameron, A. Davies, M. F. Howells and A. D. McAdam) and the dates of survey.

The maps are not published but they are available for reference at the British Geological Survey Office, Murchison House, Edinburgh, where photocopies can be purchased.

In this memoir the word 'district' means the area of land included on the Scottish 1:50 000 Geological Sheet 33E (Dunbar) and part of Sheet 41 south of the Firth of Forth.

Numbers in square brackets are National Grid references within 100 km square NT.

Numbers preceded by the letters S and ED refer to the Sliced Rock Collections of the British Geological Survey.

Numbers preceded by the letters D and MNS refer to the British Geological Survey Photograph Collection (Scotland). A full list of geological photographs taken in the district is available on application to the British Geological Survey, Murchison House, Edinburgh   EH9 3LA.

NT 55 NE	Wedder Lairs	Howells	1966–67
NT 55 SE		Howells	1967
NT 56 NE	Snawdon	Howells	1962–63
NT 56 SE	Meikle Says Law	Howells	1963–66
NT 57 NE	East Linton	McAdam	1964–67
NT 57 SE	Garvald	Howells	1963–64
NT 58 NE	Tantallon Castle	McAdam	1967
NT 58 SE	Whitekirk	McAdam	1965–67
NT 65 NW	Byrecleugh	Howells	1965–67
NT 65 NE	Longformacus	Howells	1966–67
NT 65 SW		Howells	1966
NT 65 SE	Dirrington Great Law	Howells	1966–67
NT 66 NW	Dunbar Common	Howells	1963–65
NT 66 NE	Spartleton	Davies	1965–66
NT 66 SW	Penshiel Hill	Howells	1963–65
NT 66 SE	Cranshaws	Howells	1965
NT 67 NW	Tyninghame	McAdam	1966–67
NT 67 NE	Dunbar	Davies	1964–68
NT 67 SW	Stenton	Howells	1964–65
NT 67 SE	Woodhall	Davies	1964
NT 68 SW	Lochhouses	McAdam	1966–67
NT 75 NW	Black Hill	Cameron	1962–63
NT 75 NE	Cockburn	Cameron	1962–63
NT 75 SW		Cameron	1963
NT 75 SE		Cameron	1962
NT 76 NW	Monynut Edge	Davies	1963–66
NT 76 NE	Ecclaw Hill	Davies	1963–66
NT 76 SW	Ellemford	Cameron	1961–62
NT 76 SE	Abbey St Bathans	Cameron	1962
NT 77 NW	East Barns	Davies	1964–68
NT 77 SW	Innerwick	Davies	1963–64
NT 77 SE	Bilsdean	Davies	1963

x

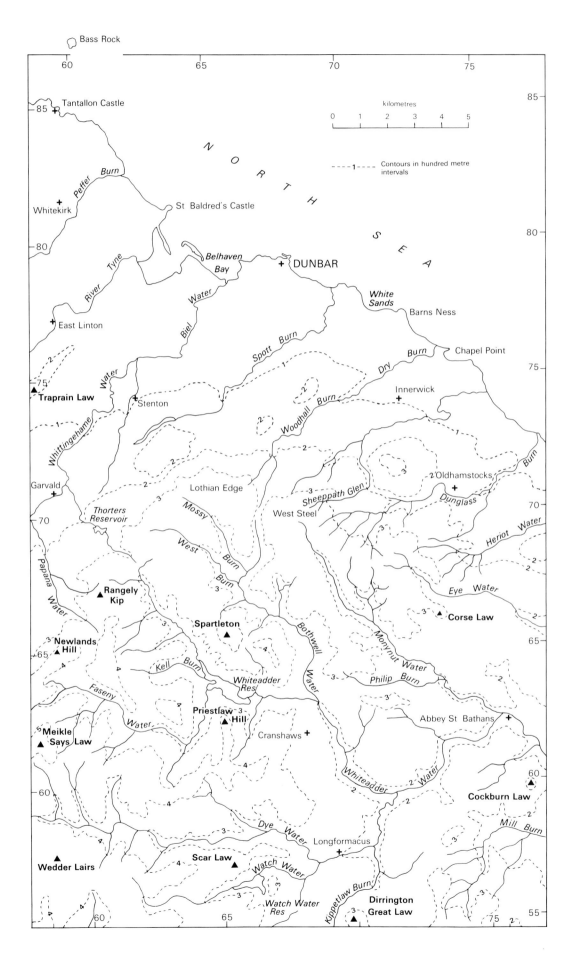

Figure 1 Main physical features of the district

CHAPTER 1

Introduction

AREA AND PHYSICAL FEATURES

This account describes the geology of the district represented on the Dunbar (33E) Sheet and a small part of the North Berwick (41) Sheet of the 1:50 000 geological map of Scotland. The district lies to the south-east of the Firth of Forth, falling for the most part within the Lothian Region, with the most southerly strip included in the Borders Region.

Topographically the district is dominated by the eastern end of the Lammermuir Hills, with Meikle Says Law (535 m) and Rangely Kipp (400 m) in the north, and Wedder Lairs (486 m) and Scar Law (361 m) in the south, being some of the higher hills (Figure 1). In marked contrast to these high, heather covered hills, utilised mainly for sheep grazing, is the arable, gently undulating ground extending north-eastward from East Linton to Dunbar, and south in a coastal strip to Oldhamstocks. Drainage of the northern area is mainly to the River Tyne, with a few small east-flowing burns to the north, such as the Peffer Burn. Drainage of the ground south from East Linton to the prominent watershed extending from Newlands Hill east to Lothian Edge, is to the Biel Water and the Spott Burn. South of this watershed the main drainage is into the River Tweed catchment via the Whiteadder Water and its principal tributaries the Faseny, Bothwell, Dye and Monynut waters. A further watershed, extending south-east from Lothian Edge through West Steel to Corse Law, separates a smaller area drained by a number of small eastward flowing rivers, notably the Eye Water, Dunglass and Thornton burns.

The markedly contrasting topography intimately reflects the underlying geology (Figure 2). The high, bleak, heather covered Lammermuir Hills are underlain by indurated and intensely folded and faulted Ordovician and Silurian greywackes and shales, and Lower Devonian greywacke conglomerate. The Lammermuir Fault, the most north-easterly extension of the Southern Upland Fault-complex, and the south-trending Innerwick Fault, separate the high ground from the gently undulating, heavily cultivated coastal area. This generally low-lying area is underlain by less resistant drift-covered sediments of Devono-Carboniferous and Carboniferous age. Carboniferous volcanic rocks, however, form the higher ground north of East Linton. Spectacular topographic features are formed by highly resistant igneous intrusions such as the phonolitic laccolith of Traprain Law, south-west of East Linton, and in the southern strip, felsitic and granitic intrusions forming respectively the prominent hills of Dirrington Great Law and Cockburn Law.

Agriculture and fishing form the main traditional industries. Coal mining on a very restricted scale was previously carried out in the Lawfield area south of Dunbar. The quarrying of limestones for burning into agricultural lime was formerly important and carried out at Oxwell Mains and along the coast at Catcraig and Skateraw Har-

bour to Torness Point. An extensive modern quarry near Dunbar exploits the thick limestones for the production of cement.

The largest centre of population is Dunbar, on the coast, with a smaller town at East Linton some 7 km to the west. A number of small villages are scattered mainly in the northern and eastern parts of the district. Detailed descriptions of the economic, geographical and historical aspects of the district can be found in the Third Statistical Account of Scotland, East Lothian volume (Snodgrass, 1953).

GEOLOGICAL HISTORY

The oldest rocks exposed are Ordovician greywackes and shales, for the most part barren of fossils, but with some graptolite-bearing shales indicating a Caradocian or possibly a Llandeilian age. Ashgillian strata may also be present. During Ordovician times great thicknesses of greywackes and shales accumulated in the NE–SW elongated Iapetus oceanic trench. The clastic proximal sediments were deposited mainly by turbidity currents, flowing transversely into, and axially along, the oceanic trench. Distal argillaceous sediments, including the graptolitic shales, are thought to have been deposited in quieter abyssal areas within the Iapetus ocean where submarine currents were generally inactive. A considerable gap in the stratigraphical record then occurs, until the deposition of Silurian strata, assigned to the Llandoverian Stage on the basis of zonal graptolites in some of the shales. These strata were deposited under similar oceanic trench conditions to those that prevailed during Ordovician times. Sedimentary structures within these Lower Palaeozoic turbidites indicate a north-west younging within NE to SW fault-bounded tracts, although the youngest strata are found in the south-east. The structure is best explained by the accretionary prism model, in which successive thin wedges of sediment are sheared from the surface of the subducting oceanic plate, underthrust beneath a stack of similar slices, and accreted at the continental margin (McKerrow and others, 1977). Some rotation of the sedimentary pile may have been caused by the underthrusting, but final rotation to the present sub-vertical attitude was probably caused by the continental collision as the Iapetus ocean closed during the climax of the Caledonian Orogeny. The resultant suture is believed now to underlie the Solway Firth.

Following this major orogenic phase, the intensely folded and faulted Ordovician and Silurian rocks formed an uplifted Southern Uplands massif. Active erosion affected much of this emergent land-mass leaving a highly irregular topography. Debris from the violent erosion was deposited during Lower Devonian times as a poorly cemented and unsorted greywacke conglomerate. Deposition of this conglomerate was generally confined to a N–S trough formed

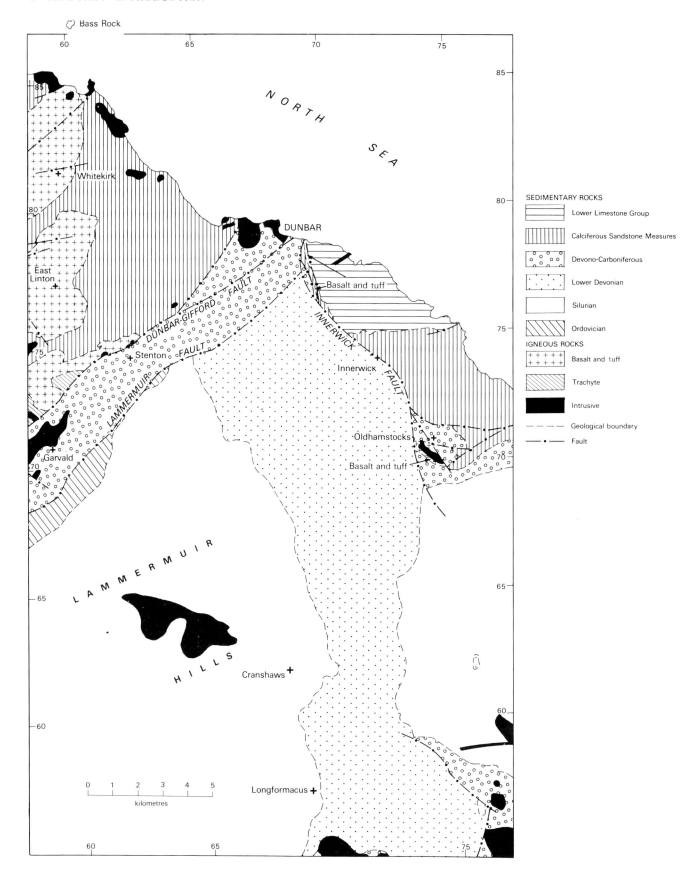

Figure 2   Generalised map of the solid geology of the district

during and immediately subsequent to the major earth movements. The original extent of this trough is unknown, but it is likely to have extended for some distance west and east of the present outcrop of the infilling conglomerate. Minor resurgences of the Caledonian Orogeny occurred during late-Silurian and possibly Lower Devonian times, accompanied by the intrusion into Ordovician and Silurian strata of the porphyrite and lamprophyre dykes and the Priestlaw and Cockburn Law granodiorites.

A more prolonged resurgence of the Caledonian Orogeny during the Middle Devonian resulted in further uplift and erosion, and no sediments of this age are preserved.

More stable conditions became established over increasingly greater areas in post Middle Devonian times. A red sandstone facies (the Upper Old Red Sandstone) deposited in fluviatile and lacustrine conditions, overlies the Lower Palaeozoic with marked angular unconformity. The lower beds of this facies in nearby areas yield a sparse fish fauna indicating an Upper Devonian age, whilst the higher beds are succeeded conformably by strata of Carboniferous age. This facies is therefore regarded as a transitional group and referred to the Devono-Carboniferous. South-east of the Lammermuir Fault thin lenticular greywacke-conglomerate, infilling hollows in the highly irregular surface of the Lower Palaeozoic rocks, forms the basal beds of the red sandstone facies.

Red and red-brown sandstones overlie the basal conglomerate and continue conformably upwards into undoubted Carboniferous strata. Intermittent periods of uplift towards the end of the red sandstone facies led to periods of soil formation in some places, now represented by irregular lenticular bands of cornstone. To the north-west of the Lammermuir Fault the basal red sandstone facies is not exposed due to faulting, and because of the absence of marker bands the relationships of the higher beds within the facies are unknown.

Conformably overlying the red sandstone development are strata of cementstone-mudstone facies referred to the lower part of the Calciferous Sandstone Measures, the lowest group included in the Carboniferous system. Deposition of similar beds probably occurred over a much greater area than that occupied by the present outcrop. Much lateral and vertical variation occurs in these sediments, reflecting depositional environments ranging from quiet shallow-water, with cementstone precipitation, to a fluviatile environment in which sandstones were deposited. In addition some irregular variation from delta-swamp to quasi-marine conditions are indicated in parts of the sequence by thin coaly beds and marine bands.

To the north-west of the Lammermuir Fault the cementstone-mudstone facies is considerably thicker, with, in general, less arenaceous material. Sedimentation was, however, interrupted by the prolonged volcanic activity which produced the lavas of the Garleton Hills Volcanic Rocks, accompanied by the intrusion of several thick sills, plugs and the Traprain Law Laccolith. Cryptovolcanic necks and associated intrusions and volcanic breccias seen westwards along the coast from Dunbar are probably also remnants of this volcanic episode. An equivalent lava-pile has not been recognised south-east of the Lammermuir Fault. Igneous activity, however, led to the intrusions at Dir-

rington Great Law, Harelawcraigs and near Plenderleith Hill, the tuff, lava and sill at Oldhamstocks, and the sill, tuff and small neck at Fluke Dub south-east of Dunbar.

Towards the end of Calciferous Sandstone Measures times, more prolonged periods of marine encroachment led to the formation of the Macgregor Marine Bands and, at the top of the group, of the richly fossiliferous Lower and Middle Longcraig limestones. Marine conditions prevailed over a considerable area as these limestones can be correlated with limestones in the Midland Valley of Scotland and in the Northumbrian Trough.

A widespread marine incursion, during which the Upper Longcraig Limestone was deposited, marks the base of the Lower Limestone Group. Rhythmic variation from delta swamp to marine conditions persisted for much of Lower Limestone Group times. Periods of partial emergence with the establishment of vegetation led to the formation of thin coal seams, usually occurring beneath the marine limestones. The higher strata of Lower Limestone Group age show increasingly less marine limestone deposition, with attendant increase in fluviatile and deltaic deposits.

Some doubt exists over the age of the youngest beds exposed in the area. A thin development of strata in the Fluke Dub area, apparently overlying the supposed lateral equivalent of the Barns Ness Limestone, may be younger than the Lower Limestone Group. Younger strata are probably present in near offshore areas.

The youngest rocks occurring in the area are west–east dykes of quartz-dolerite intruded during late Carboniferous times. These dykes may be associated with earth movements of the Hercynian Orogeny, intermittent throughout the Carboniferous and increasingly active towards the end of Carboniferous sedimentation.

Between the Hercynian Orogeny and the Pleistocene no definite evidence of the geological history of the district is known. It is likely, however, that the area during this great interval of time was mainly an emergent massif, subject to continual erosion, which has removed any younger strata that may have been deposited. It is probable that the present outcrop pattern was established by Tertiary times.

Several periods of glaciation, separated by temperate interglacial periods, occurred during the Pleistocene. The general topographical characters of the district were modified by the encroachment of successive ice-sheets from a generally westerly direction. The erosion caused by the moving ice particularly affected the softer Upper Palaeozoic rocks. A ground moraine of till was laid down blanketing most of the area, and is especially thick in the pre-glacial valleys. All of the existing boulder clay, however, probably relates to the last major (Devensian) glaciation as successive ice-sheets removed the deposits from preceding ones. Following this last glaciation, a period of periglacial conditions ensued during which thick solifluction deposits formed. Prominent channels were cut by large quantities of meltwater released from the decaying ice-sheets along the northern and eastern faces of the Lammermuir Hills, and extensive mounds and terraces of sand and gravel were deposited.

The formation of an ice-cap in the northern hemisphere during the Pleistocene caused an eustatic fall in sea-level, but in the district, relative sea-level rose as a result of isostatic

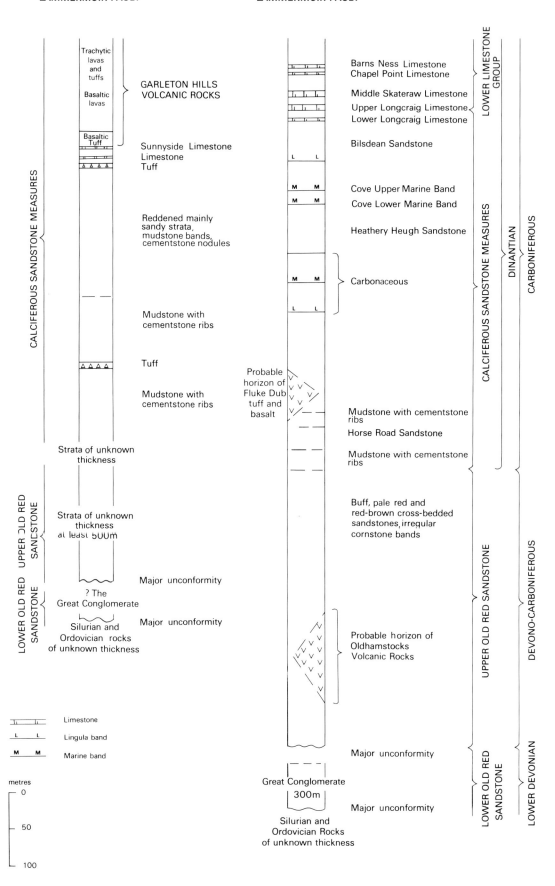

NORTH-WEST OF LAMMERMUIR FAULT

SOUTH-EAST OF LAMMERMUIR FAULT

Trachytic lavas and tuffs

Basaltic lavas

GARLETON HILLS VOLCANIC ROCKS

Basaltic Tuff

Sunnyside Limestone
Limestone
Tuff

CALCIFEROUS SANDSTONE MEASURES

Reddened mainly sandy strata, mudstone bands, cementstone nodules

Mudstone with cementstone ribs

Tuff

Mudstone with cementstone ribs

Strata of unknown thickness

UPPER OLD RED SANDSTONE

Strata of unknown thickness at least 500m

LOWER OLD RED SANDSTONE

Major unconformity

? The Great Conglomerate

Major unconformity

Silurian and Ordovician rocks of unknown thickness

Barns Ness Limestone
Chapel Point Limestone

Middle Skateraw Limestone
Upper Longcraig Limestone
Lower Longcraig Limestone

Bilsdean Sandstone

Cove Upper Marine Band
Cove Lower Marine Band

Heathery Heugh Sandstone

Carbonaceous

Probable horizon of Fluke Dub tuff and basalt

Mudstone with cementstone ribs

Horse Road Sandstone

Mudstone with cementstone ribs

Buff, pale red and red-brown cross-bedded sandstones, irregular cornstone bands

Probable horizon of Oldhamstocks Volcanic Rocks

Major unconformity

Great Conglomerate
300m

Major unconformity

Silurian and Ordovician Rocks of unknown thickness

LOWER LIMESTONE GROUP

CALCIFEROUS SANDSTONE MEASURES

UPPER OLD RED SANDSTONE

LOWER OLD RED SANDSTONE

DINANTIAN

CARBONIFEROUS

DEVONO-CARBONIFEROUS

LOWER DEVONIAN

Limestone

L   L   Lingula band

M   M   Marine band

metres
— 0

— 50

— 100

Figure 3
Generalised sections of the sedimentary and volcanic rocks in the district

depression of the land due to the weight of the ice-sheet. The maximum rise in sea-level, to about 35 m above present OD, took place in late-Glacial times. Deposits reflecting this sea-level form terraces of marine sediments which dip gently eastwards. Lower terraces at approximately 25 and 15 m above OD were formed when relative sea-level fell as a result of the isostatic uplift of the land as the ice-sheet melted. Evidence that sea-level was depressed to well below the present level occurs below low water mark, where rock platforms indicate ancient wave-cut platforms. In post-Glacial (Flandrian) times a further rise in sea-level occurred to about 8 m above OD. This level apparently remained constant for some considerable time, resulting in wave-cut platforms and extensive deposits of shelly, littoral sand and shingle. In places these deposits are backed by cliffs, and in Belhaven Bay they form an extensive flat extending inland to Preston. Subsequent fall in sea-level has resulted in the formation of extensive areas of dunes behind the present beaches, particularly from Belhaven Bay northwards to Peffer Sands. Inland peat mosses flourished in the milder post-Glacial conditions. Flat spreads of lacustrine deposits and peat mark the position of numerous small lakes. River valleys have gradually matured with the formation of terrace and flood plain alluvium.

AD

CHAPTER 2

Ordovician and Silurian

INTRODUCTION

The outcrop of the Ordovician and Silurian strata occurs in the southern part of the district in two main areas (Figure 4). These areas are separated from each other by Lower Devonian conglomerate which occupies a broad north–south tract extending from south of Dunbar to the southern edge of the district. The western outcrop of the Lower Palaeozoic rocks is bounded to the north by the Lammermuir Fault and to the east it is unconformably overlain by the Lower Devonian. The eastern outcrop is unconformably overlain by the Lower Devonian to the west and by Devono-Carboniferous rocks to the north and south.

The rocks are exposed, to a variable degree, in the rivers and streams which dissect the relatively high ground formed by the Lower Palaeozoic outcrop, but exposure between the river and stream sections is very poor, despite the absence of glacial till from large areas. This lack of exposure and the smooth rounded appearance of the hills is due to the presence of a layer of broken rock debris, which is probably the result of periglacial processes.

Rocks of Silurian age occupy most of the outcrop of Lower Palaeozoic rocks, with a small area of Ordovician rocks adjacent to the Lammermuir Fault at the western edge of the district. The position of the boundary between the two systems on the latest edition of the sheet differs from previous editions in that it is placed about 1 km farther north. The boundary is drawn on faunal evidence alone, there being no lithological basis of separation between the Ordovician and Silurian rocks.

The strata consist mainly of alternations of greywackes and shales with some occurrences of chert and black shales in the Ordovician rocks. The general lithological aspect of the Lower Palaeozoic rocks is so uniform that no lithological subdivision of the strata was made during the mapping. The strike is generally north-east to south-west, but in the vicinity of the outcrop of Lower Devonian which separates the two areas of Lower Palaeozoic rocks, the strike swings north into an orientation about 15° east of north. The dip of the rocks is normally steep, in the range from 60° to vertical, usually towards the north-west with relatively few observations of dips to the south-east. Almost invariably the direction of younging is the same as the direction of dip.

The only fossils collected from the Lower Palaeozoic rocks are graptolites and they are rather rare and usually poorly preserved. In the Ordovician the black shales usually contain some graptolites, but in the Silurian the fossiliferous lithologies are less easily distinguished. Although there are a few instances in which the graptolites identify individual stratigraphical zones, they commonly only give a more general indication of the age of the strata. The overall pattern is one in which the oldest rocks are in the north-west of the area, becoming younger towards the south-east.

LITHOLOGY

The Lower Palaeozoic rocks, particularly the Silurian rocks, show very little variation throughout the area. The sequence is built up of a series of turbidite deposits consisting of beds of greywacke which pass up into siltstone and shale. In the Ordovician rocks the turbidites are intercalated with beds of black shale and chert. No volcanic rocks have been recorded from the sequence in this area.

The beds of greywacke range in thickness from a few centimetres up to about 10 m but are usually less than 1.5 m. They normally have a sharp, sometimes erosive, contact with the underlying bed, but the top of the greywacke bed may show a more or less rapid gradation into siltstone or shale. Beds of shale occur up to 30 m thick. The proportions of greywacke to shale can vary greatly from one part of the sequence to another. Generally the proportion of greywacke in a section is in the range from 50 to 90 per cent, with the remainder consisting of siltstone and shale. This variation in the proportions of the component rock-types is not sufficiently persistent laterally to allow correlation between neighbouring stream sections.

Sedimentary structures, particularly sole-markings on the underside of the bedding planes at the base of individual greywacke beds, are quite common and are the most reliable criteria for determining the order of succession.

Other rock-types form a small proportion of the outcrop. Pale grey-green and black cherts, black shales and cherty shales which occur in the Ordovician strata are generally assumed to be hemipelagic deposits intercalated within the turbidite sequence of greywackes and shales. The sections illustrating these rock types in the Papana Water and the Thorter Burn, described by Peach and Horne (*in* Clough and others, 1910, pp. 17–19) are intensely faulted and folded. The greywackes are predominantly medium-grained, but occasionally in the basal part of the bed there may be some small pebbles. There are rare occurrences of conglomerate particularly in the Ordovician. The conglomerates consist of pebbles of chert, siltstone, shale and greywacke, loosely packed and ill-sorted in a greywacke matrix. Some pebbles are subrounded, but many, particularly those consisting of chert or shale, are angular.

STRATIGRAPHY

Peach and Horne (1899; *in* Clough and others, 1910) placed the Lower Palaeozoic strata of the Lammermuir Hills into the geographical divisions which they called the Northern Belt and the Central Belt. The Northern Belt, consisting of the Ordovician rocks, was thought to include strata of the Caradoc, Llandeilo and, in part, the Arenig series. Subdivision of the Ordovician on the 1910 edition of the geological

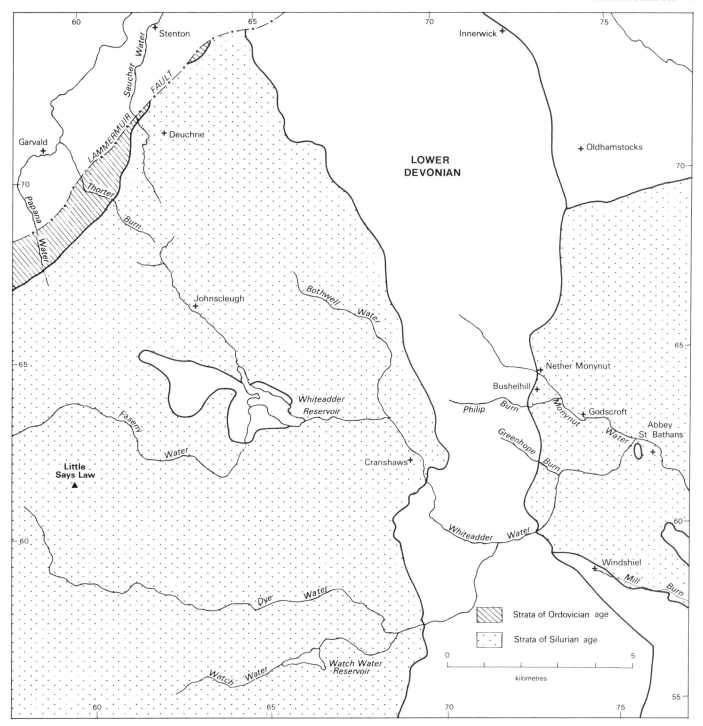

Figure 4 Generalised map of the Lower Palaeozoic rocks

map was restricted to the representation of small inliers of black shale in the Papana Water, Thorter Burn and the Sauchet Water sections. The black shales were correlated with the Glenkiln and Hartfell shales of the Moffat district by means of their graptolite faunas.

The inclusion of the Arenig Series was based on the presence of cherts, which in some sections at Ballantrae and near Abington are associated with volcanic rocks and shales containing Arenig graptolites. Subsequent work by several authors (Lamont and Lindstrom, 1957; Kelling, 1961;

Walton, 1961) has established that in other localities the association of cherts with volcanic rocks and black shales is of Llandeilian or Caradoc age.

Bergström (1971) disagreed with Lamont and Lindstrom and suggested that some of their material indicated a Llanvirnian age. Whittington and Rickards (*in* Whittington and others, 1984), however, have stated that rocks of Llanvirnian age do not appear to be known in Scotland. The oldest graptolites found during the resurvey belong to the *Nemagraptus gracilis* Zone and there is no evidence to suggest that the cherts in the Papana Water, Thorter Burn and Sauchet Water are any older.

The Ordovician rocks of this area were considered to be equivalent to the Lowther Shales by Peach and Horne and were found to contain faunas similar to the Hartfell Shales of the Moffat district.

The Silurian rocks of the Central Belt were described as Llandovery and Tarannon and no subdivision was made. They were named the Queensberry Grits and were assumed to be equivalent in age to the Birkhill Shales and the Gala Group although little faunal evidence was found.

During the resurvey it was found that lithological subdivision of the Lower Palaeozoic rocks was not practical and the boundary between the Ordovician and the Silurian was determined on graptolite evidence alone. Peach and Horne placed the boundary between the two systems at a lithological boundary which was drawn about 1 km to the south of the current position of the palaeontological boundary. The line chosen was the boundary between the Lowther Shales and the Queensberry Grits, but the graptolite evidence collected during the resurvey indicates that Silurian rocks occur to the north of this line. This may be due to structural repetition, but there are insufficient structural or palaeontological data to confirm or refute this possibility, and the coincidence of the lithological boundary with the system boundary has not been established.

In the course of the resurvey the Ordovician rocks were found to contain graptolites from the *N. gracilis* Zone and from the Lower Hartfell Shales, but the presence of the Upper Hartfell Shales, recorded by Peach and Horne, has not been confirmed. Since no evidence for the age of the cherts has been found in this area, they are assumed to be Caradoc or possibly Llandeilo in age by reason of this association with graptolite shales. In the Silurian the Birkhill Shales and the lower part of the Gala Group are represented, and the occurrence of several of the graptolite zones has been established. Faunas indicative of the *Pristiograptus cyphus*, *P. gregarius*, *Monograptus convolutus*, *Rastrites maximus*, *M. turriculatus* and *M. crispus* zones have been collected.

In general terms the faunal evidence indicates that the beds contain successively younger faunas towards the south-east, but the information is too sparse and insufficiently precise to be of use in the elucidation of the structure. Also, the fact that there are some zones which are not represented in the collection made during the resurvey has no stratigraphical significance.

STRUCTURE

According to Lapworth (1889) the structure of the Lower Palaeozoic rocks of the Southern Uplands consisted of numerous isoclinal folds which had been refolded to form a broad open anticlinorium and synclinorium. The anticlinorium, called the Leadhills Endocline, lay to the north-west of the synclinorium which was called the Hawick Exocline. Peach and Horne (1899) accepted Lapworth's interpretation of the structure and extended it throughout the outcrop of the Lower Palaeozoic rocks.

In the present district the Ordovician rocks are exposed in what was supposed to be the core of the anticlinorium, and

the Silurian rocks occupied the south-east limb of the major fold.

Lapworth's views on the structure of the Lower Palaeozoic rocks were abandoned largely as a result of advances in sedimentological research. It was found that sedimentary structures (Plate 1), which are a feature of turbidite deposition, could be used to determine the direction of younging at individual exposures. This affords direct evidence of the order of succession rather than relying solely on evidence inferred from the age of the fossils. From the period of research that followed, a general structural pattern emerged which was described by Craig and Walton (1959) and Walton (1961). The general pattern is interpreted as one of belts of repeatedly folded strata disposed alternately with belts of largely unfolded strata with a steep dip to the north-west. In the belts of folded strata, the line tangential to the crests of the folds is believed to be approximately horizontal. This general horizontal disposition of the folded belts and the NW-dipping belts taken together are equivalent in effect to a large-scale monocline. For the fossil evidence to be compatible with this structural pattern it is necessary to postulate the existence of large strike faults which bring older rocks in the north against younger rocks to the south. More recent research has led to the view that the Lower Palaeozoic rocks of the Southern Uplands constitute an accretionary prism (McKerrow and others, 1977; Leggett and others, 1979). The turbidite sequence was deposited in the Iapetus Ocean at the southern margin of the North American continent where oceanic crust was being subducted. The model involves the emplacement of successively younger packets of strata against the continental margin. Each packet was thrust under the preceding packet and was younger than its northern neighbour. Their concomitant imbrication and subsequent steepening is also due to the convergence of the crustal plates. Each imbricate wedge is bounded by a high angle thrust fault parallel to the strike.

During the resurvey it was found that detailed structural work was impractical over most of the area of the Lower Palaeozoic outcrop. The degree of exposure is generally very poor indeed and is, at best, discontinuous in stream sections. Commonly observations were limited by the degree of exposure to measurements of dip and strike. In a small number of better exposed sections the direction of younging could be determined and in only a few instances were the beds found to be inverted.

In parts of the Monynut Water and Whiteadder Water sections it was possible to delineate zones in which there are few, if any, folds and also zones in which there are many reversals of younging direction (Figure 5). This structural pattern is similar to that described by Walton (1961) although none of the postulated reverse strike faults was located. The ubiquitous occurrence of isoclinal folding as envisaged by Lapworth (1889) and Peach and Horne (1889) was not found.

DETAILS

The strike of the strata is mainly north-east to south-west and the dips are usually in the range 60° to vertical either to the north-west or to the south-east. The strike swings around from the NE–SW

Plate 1 Flute-casts in bottom surface of inverted bed of Silurian greywacke; direction of flow from top left to bottom right; quarry near Cranshaws (D 2585)

alignment close to a north–south direction in areas on either side of the Devonian belt which divides the outcrop. The strike around Cranshaws and the lower part of the Bothwell Water is about north–south. A similar strike is found on the opposite side of the Devonian outcrop at Nether Monynut [7280 6450] and Bushelhill [7270 6385], farther south in the Greenhope Burn [7330 6150] and also in the Mill Burn from about Windshiel [7450 5870] downstream to Millburn Bridge [7650 5790].

The rocks everywhere are strongly jointed and in most exposures there are signs of shearing and faulting. Cleavage is not generally developed, but it occurs locally and is fairly common in the hinge zones of the folds. The hinges of the folds are often ruptured but it is usually not possible to discover the direction or amount of movement.

The rocks close to the northern edge of the outcrop are more strongly deformed than the younger strata farther south. The Ordovician rocks which crop out in a band on the south side of the Lammermuir Fault are exposed in the Papana Water [587 682] and the Thorter Burn [603 698] near the west edge of the district. In these sections there is strong evidence of faulting and shearing over a distance of about 1 km across the strike south of the Lammermuir Fault. The rocks are steeply dipping to the north-west or are vertical and the direction of younging, where it could be determined, is to the north-west. Correlation between adjacent sections was not possible and the sections are interpreted as a series of fault slices.

The Silurian rocks farther south are less affected by faulting and the structural pattern can be illustrated by reference to sections in the Monynut Water and the Whiteadder Water around Abbey St Bathans (Figure 5). The disposition of the strata can be compared with the general structure described by Walton (1961).

The strata exposed in the Greenhope Burn [733 615] from the eastern edge of the Devonian outcrop downstream to the White-adder Water consist of a series of greywackes and shales which consistently young towards the north-west. The dip varies from about 40° to the NW to vertical and in a few instances the strata are overturned. The distance across the strike is about 1 km. This zone cannot be traced along the strike to the south-west because of lack of exposure, but a similar zone is found along the strike in the

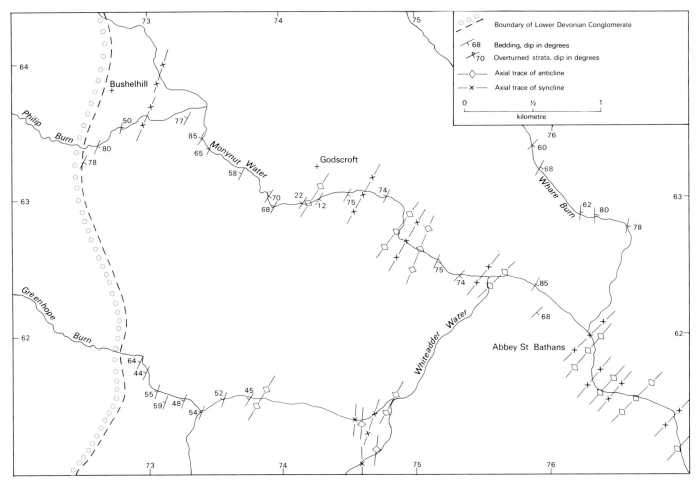

Figure 5 Disposition of axial traces in the Abbey St Bathans area

Monynut Water section from about Godscroft [7425 6320] upstream to the confluence with the Philip Burn near Bushelhill [7276 6385].

Folded strata can be seen on either side of this uniformly dipping belt in the Monynut Water and to the south-east of it in the Whiteadder Water. The thickness of strata involved in the Greenhope section is about 750 m, in the Monynut Water section it is about 1000 m. No allowance has been included for any possible movement on faults parallel to the strike. Cleavage occurs only locally. It is subparallel to the strike and dips at high angles to the north-west.

The belts of folded strata can be illustrated by reference to the section in the Monynut Water from Godscroft downstream. Part of a belt of folded strata is seen in discontinuous exposure between Godscroft and the eastern edge of the district. In this section which is about 3.5 km across strike there are at least sixteen folds.

The dip of the strata is mainly in the range 60° to vertical, and in most cases determination of younging direction indicated that the rocks are seldom inverted. Most of the folds are tight and more or less upright V-shaped folds, but in two instances the closures of the folds are much broader and open. In these cases the strata involved in the folds at the present level of exposure are thick beds of greywacke with thin partings of shale. The folds plunge at low angles to the south-west. The distance between adjacent axial plane traces varies from 50 to 350 m. Near-vertical cleavage was seen in a few instances in the hinge zone of the folds.

The exposure is inadequate to allow correlation of folded zones between adjacent stream sections along strike.

The sheet dip or *faltenspiegel* falls towards the south-east so that if strike-faulting is discounted, the rocks on the north-west side of the folded belt are lower in the succession than those exposed on the south-east side. The major structure arising out of the juxtaposition of the uniformly dipping strata and the folded belt is an assymetric anticline.   IBC

PALAEONTOLOGY

Ordovician faunas are found in the north-west part of the Lower Palaeozoic outcrop of the district. Both Glenkiln and Hartfell Shale equivalents are found as black, sometimes cherty, shales, but in many cases the fauna is poorly preserved owing to strong deformation of the rock. Most of the outcrops are close to the presumed line of the Lammermuir Fault.

Although some of the localities are closely spaced, it has not been possible to recognise a sequence of graptolite zones through any of the sections. This is not surprising since the structures are clearly complex as can be seen from the description of the section in the Papana Water given by Peach and Horne (1899) and repeated in Clough and others (1910). The faunas indicate the presence of the *N. gracilis* Zone and probably the immediately higher zones in the Lower Hartfell Shales. Peach and Horne (1899, p. 274) mentioned a locality near Deuchrie [622 710] which was said to yield an Upper Hartfell fauna. Re-examination of the material has not confirmed this view and the specimens look more like Silurian material than Ordovician. Just west of

Deuchrie, another locality has also provided a fauna which appears to be basal Silurian close to one with a Lower Hartfell fauna. There is no other evidence for Upper Hartfell in this district although there are several localities with such faunas around Lammer Law in the Haddington district to the west.

In the remainder of the Lower Palaeozoic area there are scattered localities with Silurian faunas which are broadly grouped into two belts. The one to the north-west lies around the headwaters of the Whiteadder Water and the faunas are characteristic of the Birkhill Shales (Lower and Middle Llandovery) although not all the zones seen at Moffat can be recognised. From those which can be identified, there is a suggestion that the faunas become younger towards the south-east. Similar faunas, however, occur at distances of more than 1 km across the strike and it is probable that the graptolitic horizons are repeated structurally. Most of the localities have so far yielded only a few specimens each, so that zonal interpretation is difficult since many of the species range through more than one zone. The lowest zones of the Birkhill Shales have a fauna which is largely diplograptid in composition and the *P. cyphus* Zone is the earliest which can be definitely recognised at present. It is found both in the Faseny Water [58 63] and south-east of Deuchrie [6303 7108]. The succeeding *P. gregarius* Zone is found south-east of the previous localities at a number of points between Little Says Law [59 61] and Johnscleugh [63 66]. A little further to the south-east again, a locality near the head of the Dye Water [5839 5890] has yielded a fauna suggestive of the *M. convolutus* Zone.

There is a gap of 5 or 6 km across the strike before the second belt of localities is reached. These have higher faunas of Upper Llandovery age, although mostly only slightly younger than the Birkhill faunas. The localities are more scattered and form a much broader belt with marked repetition of faunas so that no regional stratigraphy is apparent.

There is no evidence here for the lowest of the Upper Llandovery zones (*M. sedgwickii* Zone) although it may occur further east, around Siccar Point (Sheet 34). The *Rastrites maximus* Zone (which occurs in the Moffat area amongst the lowest greywackes of the Gala Group) is found in a number of places but the succeeding *M. turriculatus* Zone is not easily recognised since there is a general absence of the named species. Other species such as *Monograptus pertinax* Elles and Wood and *Rastrites distans* Lapworth occur which are normally common in the *M. turriculatus* Zone while *Monograptus exiguus* (Nicholson) is also found in quite a few localities. This last species does not occur in this area along with *Monograptus crispus* Lapworth although the two species are recorded together elsewhere. *M. exiguus* was noted by Rickards (1976) as commonest in the *crispus* Zone, although it occurs above and below this zone, while *M. crispus* itself is said to be confined to its own zone. *Monograptus discus* Tornquist is also noted by Rickards to be common in the *M. crispus* Zone. It is very rare in this area although it occurs in a number of localities further east on Sheet 34 where, however, it appears to represent a higher horizon in which it is not associated with *M. crispus*.

There are some localities which have a fauna consisting only of long ranging forms and at present these cannot be assigned to a particular zone since not even relative abundance at different horizons can be relied on. The occurrence as separate faunas in this area of species, which elsewhere in Europe usually occur together, suggests relatively rapid deposition of the strata with each graptolitic horizon representing the preservation of the fauna in the area at the time of sedimentation rather than the condensed succession characteristic of the true graptolitic facies. Under these conditions, the apparent repetition of faunas need not have any structural significance so far as the Gala Group (Upper Llandovery) is concerned and adds to the difficulty of unravelling the general structure of the area. IS

CHAPTER 3

Lower Devonian (the Great Conglomerate)

INTRODUCTION

Strata here called the Great Conglomerate form an outcrop (Figure 6) which occupies an elongated north–south tract, over 7 km wide in places, extending from just south of Dunbar to the southern margin of the district and beyond. The outcrop area is not distinguished by any topographical feature, the general form and height of the ground being similar to the flanking hills of Lower Palaeozoic greywackes and shales. The conglomerate has an estimated maximum thickness of c.350 m. It is predominantly composed of greywacke, and rests with marked angular unconformity on steeply dipping to vertical greywackes and shales of the Lower Palaeozoic.

AGE

The Great Conglomerate was formerly placed in the Upper Old Red Sandstone, and referred to as the Lower Conglomerates by Geikie (in Howell and others, 1866, p. 20) and as the great conglomerates by Bailey (in Clough and others, 1910, p. 28). There are two developments of conglomerates in the district between the Lower Palaeozoic rocks and strata of undoubted Carboniferous age. The first is the Great Conglomerate and the second is a relatively thin lenticular development, rarely exceeding 5 m in thickness, which passes upwards into red sandstones, which are in turn conformably overlain by undoubted Carboniferous strata. This latter development was also previously referred to the Upper Old Red Sandstone but is now placed in the Devono-Carboniferous (p. 16). The Great Conglomerate is now thought to be considerably older than the thin lenticular development, and is regarded as being of Lower Devonian age. This conclusion is supported by the following points:

1 The Great Conglomerate is markedly coarser-grained with, in general, a much less sandy matrix than the lenticular, basal Devono-Carboniferous conglomerate. The Great Conglomerate ranges up to probably some 350 m in thickness, whereas the latter development has a maximum thickness of about 12 m, and is conformably overlain by red-brown sandstones.

2 The brecciation of the clasts in the Great Conglomerate is a feature absent from the basal Devono-Carboniferous conglomerate. This character is here explained by a period of earth-movements which affected the former beds only.

3 A suite of north-westerly trending dykes (p. 51) is peculiarly confined to the Great Conglomerate.

4 At Bell Hill, St Abbs, to the east of the district, a lamprophyre dyke cutting a presumed outlier of the Great Conglomerate, has been dated (K:Ar) as 400 ± 9 Ma which implies a Lower Devonian age (Rock and Rundle, in press).

No diagnostic fossils have been found in the Great Conglomerate. Where it is exposed, the relationship to the underlying Silurian strata is one of marked unconformity, so that the conglomerate is definitely post-Llandovery in age. Relationships with other strata are generally faulted or obscure. Nowhere in the district has a conformable passage from the Great Conglomerate upwards into undoubted Lower Carboniferous beds been recorded. To the south of the district, west of Duns, a conglomerate, thought to be the equivalent of the lenticular Devono-Carboniferous conglomerate, overlies with apparent unconformity, strata referred to the Great Conglomerate and passes conformably upwards into fossiliferous Lower Carboniferous strata.

The brecciation affecting the Great Conglomerate probably occurred in Middle Devonian times when the district was subjected to a resurgence of the Caledonian earth-movements. Igneous activity, perhaps contemporaneous with, although possibly later than these earth-movements, gave rise to the intrusion of the north-westerly trending suite of dykes which cut the Great Conglomerate but not the Devono-Carboniferous rocks. This placing of the Great Conglomerate in the Lower Devonian conforms with the thick developments of Lower Devonian conglomerates in other areas of Scotland.

DETAILS

The conglomerate is composed mainly of pebbles, cobbles and boulders of greywacke set in a generally loosely cemented matrix of similar finer material, ranging in colour from red- to purple-brown. The cement is most commonly weak and ferruginous, but in some areas, notably in Sheeppath Glen [6950 7040] and Back Water [6994 6997], a carbonate cement occurs, resulting in a more cohesive conglomerate. This variation in cement has resulted in spectacular 'badlands' weathering along the Oldhamstocks Burn and tributaries [7010 6995]. Boulders up to 0.7 m are fairly common, but in general cobbles up to 0.31 m are more characteristic. The dominant greywacke clast content is leavened in places by scattered pebbles and cobbles of quartzite and badly weathered dioritic rock. Marked cross-bedding occurs at many localities emphasised by the alignment of the long axes of the constituent clasts. A noticeable brecciation of these clasts is characteristic. Lenticular cross-bedded rafts of red-brown sandstone, up to 6 m thick, occur in places, but appear to die out rapidly laterally. Thinner bands of sandstone and silty sandstone, in places with pebbles, occur throughout the conglomerate.

The Great Conglomerate was deposited on an irregular surface of steeply dipping Silurian greywackes and shales, within the confines of an elongated north–south trough. The original extent of this depositional trough is unknown, but was certainly over a wider area than the present outcrop. In the Hartside Edge area [7680 7135] the extremely uneven nature of the pre-Devonian topography is well illustrated. Here the feathered edge of the present outcrop limit is well exposed, with 'windows' of greywacke and shale, particularly well seen in Dean Burn [6572 7184]. Farther south along the western margin of the tract, the relationships between Devonian and Silurian strata are obscure due to poor exposure. The general trend of this western margin of the conglomerate is however north-north-west to south-south-east, but in detail variable, suggesting an

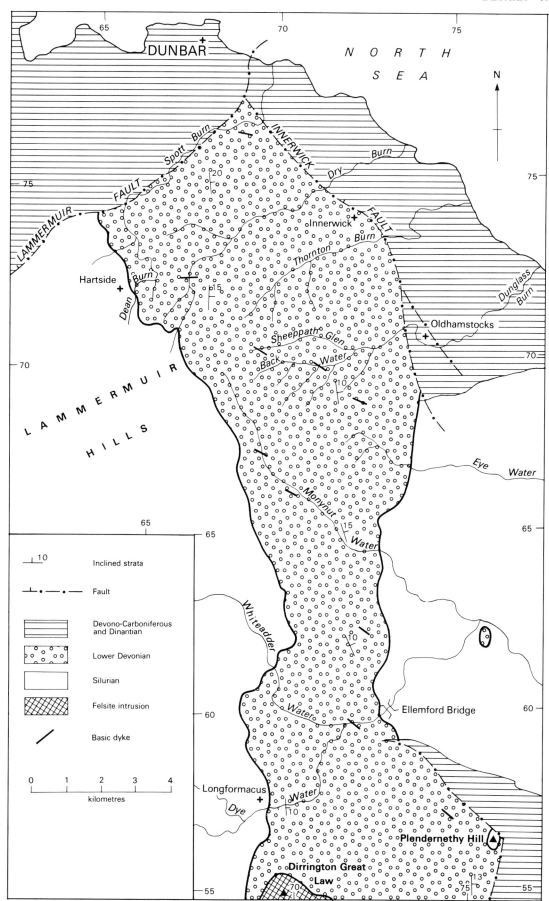

Figure 6
Generalised map of
the Lower Devonian
rocks

Plate 2   Great Conglomerate, Lower Devonian, showing range of grades and fracturing of the boulders and slight imbrication, Burn Hope near Oldhamstocks (D 2745)

Plate 3   Valley cut in the Great Conglomerate, Lower Devonian, showing recent erosion features, Back Water (D 3661)

unconformable and not a faulted junction. One exposure in the Thorter Cleugh [6694 7029] does show the conglomerate faulted against greywackes and shales. This fault position is, however, occupied by an acid porphyrite dyke, and cannot be traced outwith the immediate exposure. South of the Thorter Cleugh although the exposures are generally poor, the weight of evidence strongly indicates an unconformable junction.

The north-western boundary is formed by the Lammermuir Fault, which downthrows Devono-Carboniferous strata to the north-west against the Great Conglomerate. It is possible that remnants of the Great Conglomerate occur to the north-west of the Lammermuir Fault, but no exposures or indications are known at present of the original northwards extent of the depositional trough. For some 9.5 km southwards from Dunbar to near Oldhamstocks, the eastern boundary is formed by the Innerwick Fault, which effectively downthrows Carboniferous and Devono-Carboniferous strata to the east against the Great Conglomerate. South of the supposed termination of the Innerwick Fault the junction is poorly exposed. One exposure occurs in the Eye Water [7354 6677] where a basal conglomerate, rather brecciated with angular to subangular greywacke and shale clasts set in a calcareous-cemented, silty, red-brown matrix, overlies shales and greywackes dipping at 60° to the west. No further exposures are seen of this basal junction, but the general indications are of a steep angular, unconformable relationship with the Silurian greywackes and shales.

Approximately from 0.75 km south of Ellemford Bridge [7310 5915] south-eastwards to Plendernethy Hill [7590 5640] the Great Conglomerate–Devono-Carboniferous junction is faulted, with a further fault forming the junction from Plendernethy Hill to the margin of the district at [7550 5455]. A number of basaltic dykes, trending within the arc W5° to N35°W, cut the Great Conglomerate. Dykes of this trend are peculiar to the outcrop of the conglomerate, and are thought to be of possible Carboniferous age (p. 51).

AD

CHAPTER 4

Devono-Carboniferous (Upper Old Red Sandstone)

INTRODUCTION

The strata here classified as Devono-Carboniferous were formerly placed in the Upper Old Red Sandstone and their age was assumed to be Upper Devonian. They pass upwards, with no apparent unconformity, into undoubted Lower Carboniferous sediments and fossil evidence from miospores suggests that the base of the Carboniferous lies within the sequence formerly classified as Upper Old Red Sandstone. The term Devono-Carboniferous is used as a transitional division as the Devonian–Carboniferous boundary cannot be defined.

The principal area of outcrop of these rocks in the district lies between the Dunbar–Gifford and the Lammermuir faults in a belt running south-westwards from Dunbar to the Garvald area (Figure 7). Other smaller areas are present in the vicinity of Oldhamstocks and in the south-eastern corner of the district.

Exposures of the Devono-Carboniferous strata are poor and discontinuous but the general sequence of beds, in descending order, is as follows:

4 Cross-bedded, red-brown sandstones with much hematite enrichment in nodular and vein form, and irregular hematite-stained cornstone concretions.

3 Thick sequence of cross-bedded, red-brown sandstones, some massive bands; brick-red siltstones; silty mudstones and blocky weathering mudstones with thinly-bedded sandstone ribs and bands.

2 Series of red and red-brown sandstones commonly with lenticular pebbly bands in a silty matrix, clasts angular and subangular to rounded, generally up to 0.05 m.

1 Basal conglomerate of variable thickness, 0 to 12 m.

These sediments are considered to have been deposited in a partly fluviatile and partly lacustrine environment. The conglomerates accumulated in alluvial fans which extended northwards from high ground in the south. The sandstones, siltstones and mudstones were deposited in an alluvial plain. The cornstones in the upper part of the succession are interpreted as pedogenic carbonates which were derived from the calcareous horizons in former soils.

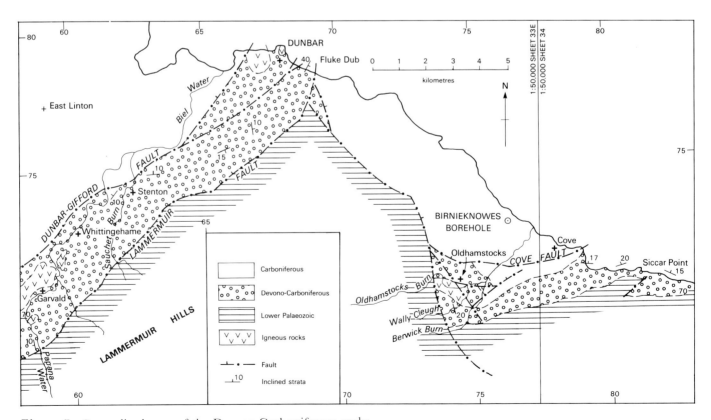

Figure 7 Generalised map of the Devono-Carboniferous rocks

AGE

The only fossils recorded from the Devono-Carboniferous strata in the district are single scales of the fish *Holoptychius nobilissimus* from a fallen block in a river gorge near Whittingehame (Bailey *in* Clough and others, 1910, p. 35) and from a depth of 398 m in the IGS Birnieknowes Borehole [7580 7317], some 65 m below the horizon taken to be the top of the Upper Old Red Sandstone facies.

An indication of the age of the upper part of the sequence can be inferred from research done on Lower Carboniferous miospores. Neves and others (1973, figs. 10, 11, 15) placed the lowest undoubted Lower Carboniferous sediments in the district, and adjacent areas, in the miospore concurrent range zone CM. These beds overlie, with no apparent unconformity, strata of Upper Old Red Sandstone facies. As the CM zone is not the basal zone of the Carboniferous in the classification based on miospores, it is reasonable to infer that part of the Upper Old Red Sandstone is of Carboniferous age. As no recognisable miospores have been recovered from the sediments of Upper Old Red Sandstone facies, it is not possible to draw the position of the Devonian – Carboniferous boundary in the district.

DETAILS

The basal beds of Devono-Carboniferous age are exposed only in the Papana Water [5858 6880] immediately to the north of the Lammermuir Fault. Here a lenticular development of conglomerate up to 12 m thick is exposed underlying cross-bedded sandstones. The conglomerate is mainly composed of greywacke clasts with rare lava pebbles, set in a pale red, sandy matrix, generally well cemented. Clasts range in shape from angular to subrounded, generally about 0.15 m in diameter, with rare examples up to 0.38 m. The precise relationships of this conglomerate to the overlying beds, and to the underlying greywackes, are not seen. A few exposures in the Pease Bay area to the east of the district indicate, however, that the conglomerate was laid down on an irregular surface and is conformably overlain by sandstones.

The sequence overlying the basal conglomerate is not well exposed, but scattered exposures show that the strata are mainly sandstones (Plate 4). Interspersed with these sandstones are finer silty bands. Green, reduction spots and ribs are features of these finer bands which are more numerous towards the middle of the development. Some channelling and cross-bedding are characteristic of much of the sequence, accompanied by numerous layers of pellets of siltstone and silty mudstone, often angular and flaky.

Eastwards from Dunbar to Fluke Dub a series of red and red-brown sandstones with occasionally silty bands is well exposed

Plate 4   Bedded Devono-Carboniferous sandstones, showing eastward dip and minor faulting, East Bay, Dunbar (D 3652)

(Plate 4). These beds are generally uniformly bedded with small-scale cross-bedding. Towards the top of this sequence, cross-bedded, red and red-brown sandstones with much hematite in nodules and veins, and irregular calcareous concretions of corn-stone, give a characteristic appearance to this upper development.

The presence of cornstones (pedogenic limestone) is restricted to the upper part of the Devono-Carboniferous, and therefore is a very useful stratigraphic indicator.

In the IGS Birnieknowes Borehole some 104 m of strata referred to the Devono-Carboniferous were recorded. These rocks are mainly red-brown, fine- to medium-grained, cross-bedded, friable sandstones. Concentration of hematite in ribs, veins and nodules is common throughout. Irony pellets in layers, and deep red-purple stained, silty mudstone fragments are also characteristic, as are the scattered pale green reduction spots and occasional green siltstone ribs. Concretionary cornstone bands, generally less than 0.6 m thick, are also characteristic. In the upper 30 m, brick-red irony ribs

are also a feature. Fish remains are scarce; indeterminate fragments only were recovered from four horizons, and one large fish plate referred to *Holoptychius nobilissimus* was found in a medium-grained, brownish-red sandstone at 463 m. The top of the Devono-Carboniferous in this borehole is arbitrarily taken beneath a thin development of cementstones, siltstones and mudstones, the base of which is at 398 m.

Exposures of strata occurring at approximately similar horizons in the upper part of the sequence are seen in Wally Cleugh [7435 6974] upstream to [7370 6941] and in the Berwick Burn [7541 7066] upstream to [7528 7038] and [7522 7009]. The sporadic nature of these outcrops is such that precise correlation from one stream to another is impossible. However, the overall stratigraphy of these beds is such that a general correlation can be made. The upper limit of this development, as drawn on the 1:50 000 Geological Sheet 33E, is taken above the cornstone development and beneath the cementstone/mudstone development. This horizon may vary in stratigraphical horizon from locality to locality.   A D

CHAPTER 5

Carboniferous

INTRODUCTION

Strata of definite Carboniferous age are confined to the northern part of the district. They are all placed in the Lower Carboniferous (Dinantian) with the possible exception of a small area which may be basal Namurian. The succession is divided into the Calciferous Sandstone Measures and the overlying, and much thinner, Lower Limestone Group.

To the north-west of the Lammermuir Fault only the lower division is present, comprising a thick sequence of sediments and the Garleton Hills Volcanic Rocks. This sequence is not well exposed but part of the succession beneath the Garleton Hills Volcanic Rocks was recorded in detail from the East Linton IGS Borehole.

To the south-east of the Lammermuir Fault the Lower Limestone Group, with thick economic limestones, is also present. In this area the Dinantian succession is reasonably well known from surface exposures, especially coastal sections, from borehole evidence, in particular the IGS boreholes at Skateraw and Birnieknowes, and from quarries. There is a marked scarcity of volcanic rocks in this area compared to that north-west of the fault.

The palaeontology and zonation of the sedimentary rocks are dealt with in Chapter 6 and the igneous rocks are described in Chapters 7 and 8.

CALCIFEROUS SANDSTONE MEASURES

As previously stated (p. 17) the base of the Carboniferous lies within the sediments here classified as Devono-Carboniferous. For practical purposes, however, the base of the Carboniferous is taken on the 1:50 000 Geological Sheet 33E at a marked lithological change, where red sandstones with cornstones are succeeded by a mudstone-cementstone facies with sandstones and siltstones, the mudstones yielding Carboniferous fossils. From the miospore zonation (Neves and others, 1973) this would appear to be some distance above the true base of the Carboniferous. This marked lithological change, here taken as the base of the Carboniferous, may be diachronous as the cornstone-sandstone and mudstone-cementstone lithologies probably interdigitate. An outstanding characteristic of the Calciferous Sandstone Measures is the rapid lateral and vertical variation of lithologies. The lateral variations, although marked, are seen to occur within groups of beds of extensive lateral extent (Figure 8). An example of this is provided by the Macgregor Marine Bands which, although highly variable in detail, can be traced from Northumberland to the Lothians and Fife (Wilson, 1974). Evidence of the lateral persistence of some of the thick sandstones in the district and beyond was supported by research on the vertical distribution of miospores in the Calciferous Sandstone Measures (Neves and others, 1973, figs. 9,11). Figure 8 gives some indication

of the distribution and variation of the lithological sequences within the Calciferous Sandstone Measures in the district and in adjoining areas.

The succession can be roughly divided into three groups in descending order:

c Mainly arenaceous strata with the Macgregor Marine Bands at the base and the marine Lower and Middle Longcraig limestones at the top.
b Strata with much carbonaceous material with impersistent marine bands.
a Strata of predominantly cementstone-mudstone facies.

Detailed correlation of individual beds within these three broad divisions is generally not possible because of the degree of lateral lithological variations in the sequences in different parts of the district. This is particularly so in the case of the cementstone-mudstone facies to the south-east of the Lammermuir Fault. The few lithological units, such as the Horse Road, Heathery Heugh and Cove Harbour sandstones, show great variations in thickness. The marine limestones at the top of the sequence are almost uniform in thickness and can be traced over considerable areas.

The sequence of sediments suggests an initial period of mainly shallow lagoonal deposition followed by deltaic and fluvial conditions with sporadic marine incursions becoming more pronounced towards the close of Calciferous Sandstone times.

To the north-west of the Lammermuir Fault the cementstone-mudstone sequence is overlain by the Garleton Hills Volcanic Rocks. To the south-east of the fault, no definite correlatives of the volcanic rocks have been recognised. In the latter area, volcanic rocks crop out at Fluke Dub [693 784] and at Oldhamstocks [735 705] but the exposures are poor and the field relationships are obscure. The Fluke Dub occurrence may be contemporaneous with the Garleton Hills Volcanic Rocks but the Oldhamstocks development may well be somewhat older (p. 36).

Cementstone-mudstone group

This basal part of the Calciferous Sandstone Measures is mainly composed of alternating cementstones (dolomitic limestones) and mudstones with sandstones and siltstones also present. Sandstones are less common to the north-west of the Dunbar–Gifford Fault than in the eastern part of the district. In the north-western area, towards the top of the group, tuffaceous sediments occur. These mark the onset of the volcanic episode and are interbedded in places with unfossiliferous, impersistent, limestone bands. The Sunnyside Limestone in the East Linton area is the best developed of these and its stratigraphical position was identified in the East Linton Borehole (Figure 11).

Exposures of the cementstone-mudstone facies are seen in coastal sections west of the Parade Vent, at Dunbar. On the east side of Belhaven Bay and eastwards towards the Parade Vent, up to 65 m of strata of this group are exposed. Cementstones, exceptionally up to 0.75 m thick, are brickred in colour and separated by mainly silty and muddy strata with a fauna mainly of *Spirorbis sp.* and bivalves, which are not stratigraphically diagnostic.   AD

Other exposures of these rocks north-west of the Dun-

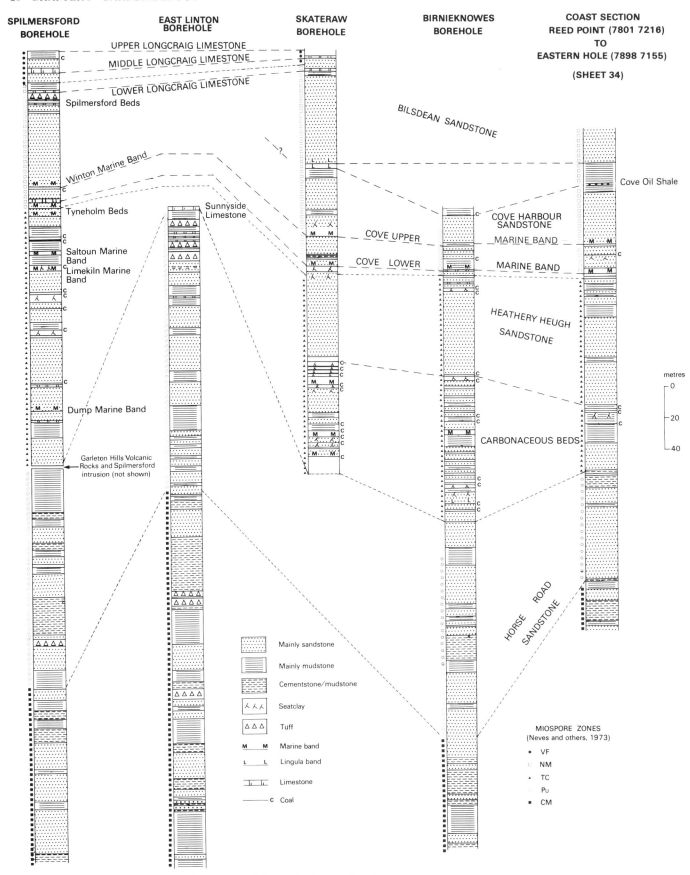

Figure 8   Comparative vertical sections of the Calciferous Sandstone Measures

bar–Gifford Fault are seen mainly on the foreshore at Tyne Mouth, Car Rocks and a discontinuous section in the Biel Water. To the north of the mouth of the River Tyne on the foreshore, about 40 m of strata are seen dipping gently to the north-west, cut by several small west-north-west to east-south-east faults. Fairly thick beds of sandstone form the base, middle and top of this sequence, separated by soft red and green mudstones with red hematite-stained cementstone ribs. On the foreshore between Car Rocks and Scoughall Rocks, about 160 to 170 m of strata are exposed, cut by a small north-east to south-west fault. To the south-east of this fault about 110 m of alternating bands of sandstone and marl are the lowest beds in the section. To the north-west of the fault, some 60 m of mainly red and grey marls and mudstones and rare sandstone bands are present. Bands of purple tuff occur in the marls and mudstones. These strata may correlate with those seen in the upper part of the East Linton Borehole (Figure 11). In the Biel Water a good, but discontinuous, section is exposed from the raised beach at Beltonford to the Dunbar–Gifford Fault near Biel. In all, about 60 m of strata dipping gently south-south-east are seen, with some further isolated exposures upstream to the confluence with the Sauchet Water. White and purple sandstones are dominant, interbedded with red and green mudstones with numerous red, hematite-stained, cementstone bands. Plant fragments and worm tracks are seen in greyer, less-stained cementstone bands. Bands of purple tuff are present in the higher parts of the section. These beds are thought to be again possibly equivalent to those seen in the upper part of the East Linton Borehole. ADM

To the east of the Parade Vent one small fault- and vent-bounded area of cementstone-mudstone facies occurs but is of insufficient extent to indicate the stratigraphical horizon. A further small area of similar sediments with a cementstone-breccia occurs about 150 m west of Fluke Dub, to the west of a faulted ridge of basalt and intrusive basanite and associated tuff which, with faulting, obscures the stratigraphical relationships.

In the eastern part of the district exposures of the cementstone-mudstone facies are confined to disjointed outcrops in the Oldhamstocks–Dunglass Burn and its tributary the Berwick Burn. In the Oldhamstocks Burn exposures, sandstones with silty and shaly bands and bright red, cementstone nodules overlie a 0.3-m thick cementstone, which in turn overlies calcareous sandstones and mudstones. Red iron-staining characterises all lithologies with some greenish reduction colours showing in the mudstones. In the Berwick Burn, sandstones are again dominant, with a thin, red-stained, cementstone-breccia also present. These two sections occur near the base of the cementstone-mudstone group. The greater part of the group is cut out downstream by the Cove Fault. Further exposures, 50 to 300 m upstream of Berwick Bridge [7509 6982], are thought to be in the lower part of the cementstone-mudstone development, the rocks being mostly buff, red and white sandstones, generally cross-bedded. Towards the base of these sandstones, about 300 m upstream of Berwick Bridge, a thin, pale yellow-brown cementstone occurs. Below this cementstone (i.e. upstream) there are no exposures for some 25 to 30 m, and as the next exposure is of brick-red cornstone, it is convenient to take the base of the Carboniferous strata here, thereby placing the

cornstone in the Devono-Carboniferous. The broad sequence of strata described above from scattered outcrops is confirmed in detail by the Birnieknowes Borehole (Figure 8). In the sequence proved, the cementstone-mudstone group can be roughly subdivided into four lithological groups: **a** basal cementstone-mudstone development, **b** Horse Road Sandstone, **c** upper cementstone-mudstone, **d** upper barren mudstone-siltstone development with sandstone bands and ribs, characterised by green and purple staining. The presence of a thick sandstone separating the cementstone developments suggests that this area south-east of the Lammermuir Fault may have been at times less stable than that to the north-west.

Carbonaceous group

Strata referred to the carbonaceous group are not exposed. Borehole records, in particular the Birnieknowes and Skateraw boreholes, indicate that the group persists, probably at least as far as the Lammermuir Fault. To the north-west of this fault these beds are unknown. The group may be roughly subdivided into a lower part with thin, impersistent, foul coals and carbonaceous mudstones, overlain by a mainly sandy development, the Heathery Heugh Sandstone. The thin coals, carbonaceous mudstones and seatclays of the lower division are separated by thin sandstone bands and two impersistent marine horizons, known only from the Birnieknowes Borehole. None of the coal seams recorded is of an economic thickness. The overlying Heathery Heugh Sandstone is pale red, medium-grained with coarser bands, strongly cross-bedded, mottled in parts with small, deep purple, hematitic concretions often aligned along bedding and cross-bedding planes.

Arenaceous group with marine strata

The uppermost part of the Calciferous Sandstone Measures is characterised by the Macgregor Marine Bands at the base, two marine limestones at the top and several thick sandstones.

Macgregor Marine Bands

The Macgregor Marine Bands (Wilson, 1974) are evidence of the first fully marine incursion to affect the district in Carboniferous times. At the type locality, the Spilmersford Borehole (Davies, 1974) to the west, three distinct bands are present. In the present district, however, only one or two bands have been recorded at any one locality.

Exposures are seen at a number of localities in the vicinity of Thornton village [738 734] (H. H. Wilson, 1952, p. 308; R. B. Wilson, 1974, p. 64). The former author cited four localities with the marine fossils occurring in different lithologies, ranging from mudstone with ironstone nodules to calcareous sandstone. He considered, however, that they could all be correlated with the Cove Lower Marine Band, the lower of two bands on the coast at Cove (Figure 8), just beyond the eastern margin of the district. Wilson (1974) correlated the various marine strata in the Thornton area and the bands at Cove with the Macgregor Marine Bands of the Spilmersford Borehole, emphasising that the sequence containing the marine bands is highly variable even when traced

Plate 5 Doo Cove Sandstone, showing contemporaneous slump structures, Bilsdean (D 1148)

over relatively short distances. The faunas of the Macgregor Marine Bands consist mainly of brachiopods and bivalves with several species confined to this part of the succession (Wilson, 1974, p. 43). Another exposure correlated with the marine bands is that of the Linkhead Limestone [7585 7355] which consists of a calcareous sandstone grading down into a sandy limestone, totalling about 4.5 m, which has yielded a rich fauna of brachiopods with corals and crinoid debris.

Representatives of the marine bands were recorded in two IGS boreholes in the district. In the first, at Skateraw [7373 7546], two bands were present. The upper one was of pale, purple-grey, silty mudstone with brachiopods and bivalves and the lower one was of grey mudstone with bivalves. In the Birnieknowes Borehole [7580 7317] a calcareous sandstone with bivalves and crinoid remains was overlain by some 5 m of grey siltstones with marine molluscs and crinoid debris. It is possible that three marine bands are represented in the above sequences, but detailed correlation is difficult because of the restricted fauna and marked lithological variation. The two marine bands in the Skateraw Borehole are the correlatives of the Cove Upper and Lower marine bands. The marine band recorded in the Birnieknowes Borehole is the equivalent of the Cove Lower Marine Band.

Above the marine bands a variable sequence of strata occurs, the finer-grained members separated by variable sandstone bands which may vary laterally from a few metres up to 20 m in thickness. One of these sandstones, the Cove Harbour Sandstone, is of variable thickness, but appears to persist over much of the Calciferous Sandstone Measures outcrop. Between the Cove Harbour Sandstone and the Bilsdean Sandstone at the top of this mainly arenaceous group, an alternating sequence of silty mudstones, siltstones and sandstones occurs. Some of these sandstones show marked lateral variation of thickness. The most notable of these can be seen from Doo Cove [7654 7290] to Otter Hole [7648 7296] (Plate 5). Here a sandstone about 2 m thick at Doo Cove thickens in about 100 m laterally to about 12 m at Otter Hole. Characterised by marked cross-bedding, this sandstone is probably a channel-infill as rapid attenuation also takes place northwards from Otter Hole. Within this alternating group of strata in the Skateraw Borehole, two *Lingula* bands separated by about 2 m of sandstone occurred at the base of the Bilsdean Sandstone. No outcrops of these bands are known.

The Bilsdean Sandstone, forming the upper part of this group, can be up to 50 m thick, is persistent, and forms a good aquifer. In general it is a fine- to medium-grained sandstone with frequent, coarse-grained, friable bands, a conglomerate band up to 1 m thick, and a few fine-grained ribs. A characteristic, faint, red staining is generally present with bands containing angular clasts of silty mudstone and mudstone.

The highest strata of the Calciferous Sandstone Measures are notable for the occurrence of two persistent and characteristic limestones both yielding a diagnostic shelly fauna. These limestones are the Lower and Middle Longcraig limestones, correlatives of which are present over most of the Midland Valley of Scotland. The Lower Longcraig Limestone is the first physical evidence of a widespread marine incursion during the Carboniferous that can be traced over most of central Scotland. Details of the contained fauna are given in Chapter 6, and by Wilson (1974).

Strata beneath the Lower Longcraig Limestone are only exposed in the railway cutting [6980 7673], (b) below, where some 5 m of mudstone and siltstone with calcareous nodules and thin limestone ribs with *Lithostrotion junceum* colonies and brachiopods underlie the limestone. This development shows that marine conditions were present over the area before the limestone was deposited.

Lower Longcraig Limestone

The Lower Longcraig Limestone varies up to 3.35 m in thickness, but averages less than 2 m. In general it is a rather argillaceous, brownish-grey limestone with bands of calcareous siltstone, and is rather nodular in most places. Brachiopods, crinoid debris and corals are the main fossils found. Three exposures of this limestone occur:

a at Longcraig [7510 7511] on the coast where the thickest development is seen of 3.35 m of brownish-grey, hard, compact limestone with calcareous siltstone partings, grading downwards into nodular limestone with brachiopods, crinoid debris and scattered colonies of *Lithostrotion junceum* in the harder limestone ribs.
b railway cutting [6980 7673], where 1.22 m of grey, unevenly bedded, fine-grained limestone with calcareous siltstone ribs and bands yields shells and crinoid debris.
c Dry Burn [7141 7498], upstream from Ford Bridge, where 1.22 m of grey, fine-grained, hard, compact, nodular limestone with crinoid remains and shells are seen.

Overlying the Lower Longcraig Limestone, fossiliferous calcareous siltstone and mudstone with thin limestone bands and nodules form the upper part of the marine cycle. These

Plate 6 Basin-shaped depressions in the upper surface of the Middle Longcraig Limestone, Catcraig (D 3654)

Plate 7   Sedimentary cycle, Middle Longcraig Limestone at base, seatclay (behind hammer shaft), thin coal (at upper end of hammer), marine mudstone (dark), upper Longcraig Limestone at top, Catcraig (D 3655)

sediments are overlain by mottled pale purple-red and buff sandstones, with large-scale cross bedding and friable bands.

Middle Longcraig Limestone

The Middle Longcraig Limestone indicates the next period of marine deposition. This limestone is sometimes known as the 'spaghetti-macaroni' rock from the appearance of the weathered surface as it contains the remains of numerous specimens of *Lithostrotion* and solitary rugose corals. This coral limestone is best exposed along the foreshore at Catcraig [7159 7633] where the original upper surface, pitted with regular depressions (Plate 6), is exposed by the erosion of the overlying, infilling, blue-grey seatclay. These hollows may be the result of a natural interference pattern of growth of the

coral colonies which was terminated by the recession of the sea or maybe a result of incipient karstification of the limestone during the period of seatearth formation (c.f. Walkden, 1974). The Middle Longcraig Limestone varies from 0.69 to 2.87 m thick, although most often about 1.5 m thick, generally persistent and characteristic with a rubbly upper part mixed with seatclay.

The blue-grey seatclay infilling the regular hollows in the roof of the Middle Longcraig Limestone is overlain by carbonaceous mudstone or a thin, foul-coal layer (Plate 7). Calcareous siltstone or mudstone yielding abundant calcareous brachiopods and crinoid debris overlies the coaly horizon, indicating a return to more marine conditions. This is the lower part of the cycle containing the Upper Longcraig Limestone, the base of which is taken as the top of the Calciferous Sandstone Measures.

LOWER LIMESTONE GROUP

The Lower Limestone Group is characterised by repeated periods of marine conditions during which fossiliferous limestones were deposited. These marine episodes were separated by periods of emergence leading to the establishment of vegetation and the subsequent formation of thin coal seams. The lower part of this group is exceptionally rich in marine invertebrate fossils. The upper part is less rich as the depositional area became less stable with an attendant increase in the supply of terrigenous material and the deposition of thick sandstones with scarce, relatively poorly fossiliferous bands of limestone.

The base of the Lower Limestone Group is taken at the base of the Upper Longcraig Limestone (Figure 9). This is the local equivalent of the Hurlet Limestone of the Glasgow area, a widespread and well-documented limestone yielding a characteristic fauna which has facilitated accurate correlation over most of the Midland Valley of Scotland.

Upper Longcraig Limestone

The Upper Longcraig Limestone ranges from 5.53 to 8.99 m thick, generally averaging some 6.96 m. A rich and varied fauna facilitates the accurate correlation of this limestone over the area, and the general lithological character remains remarkably constant. Rugose corals are characteristic, forming a noticeable band up to 0.31 m thick about 0.90 m below the top. Colonies of *Lithostrotion*, brachiopods and crinoid debris are also present in abundance throughout the limestone but particularly near the top and bottom.

Outcrops of this limestone occur at Long Craig (type locality) [7490 7513] south of Torness Point where 6 m of limestone with a well-developed, rugose coral band forms the prominent reef of Long Craig; in the Dry Burn 230 m upstream of Ford Bridge [7152 7505], where 6.55 m of limestone occurs yielding *Lithostrotion*, brachiopods and crinoid debris; and at Cat Craig where up to 7.67 m of limestone are exposed with a well-developed band of rugose corals near the top. Brachiopods and *Lithostrotion sp.* colonies are also common. The railway cutting [6983 7674] affords a further good section through this limestone where 7.8 m of muddy limestone, with a well-developed band of rugose corals 0.76 m from the top, brachiopods, crinoid debris and *Lithostrotion* colonies, are seen. The Upper Longcraig Limestone has been worked at Cat Craig, and seawards on the wavecut platform, when it was used in the production of lime. It is now being worked extensively at Dunbar (formerly Oxwellmains) Quarry for the manufacture of cement.

Strata varying from about 5.5 to 6.5 m intervene between the top of the Upper Longcraig Limestone and the Lower Skateraw Limestone. Overlying the Upper Longcraig Limestone in many places is a variable thickness of dark grey mudstone, often very fossiliferous. This mudstone is, however, cut out in some parts by the overlying sandy beds which are characteristically cross-bedded, and bioturbated with sand-filled worm-burrows, particularly in the lower part. The sandy development is overlain by seatclay which is capped by the Lower Skateraw Limestone.

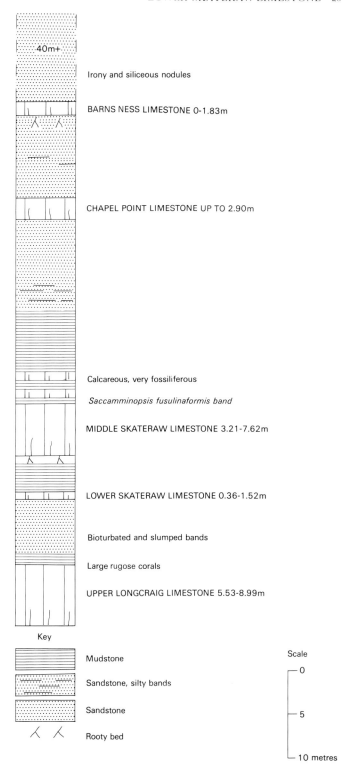

Key

Mudstone

Sandstone, silty bands

Sandstone

Rooty bed

Scale

0

5

10 metres

Figure 9 Generalised vertical section of the Lower Limestone Group

Lower Skateraw Limestone

The Lower Skateraw Limestone ranges from 0.36 to 1.52 m in thickness, and is generally a compact, brownish-grey, dolomitic limestone characterised by large productoid brachiopods. In the Dunbar quarry this limestone is discarded as its higher magnesium carbonate content makes it unsuitable for cement manufacture. Outcrops of this limestone occur at similar localities to those detailed for the Upper Longcraig Limestone. The limestone is mainly interesting for its persistent uniform character. Exposures are seen at the following localities:

a On the coast between Torness Point and Longcraig [7488 7517], 0.84 m of brownish-grey limestone with scattered large productoids, colonies of *Lithostrotion* near the top, a few rugose corals and crinoid debris.
b Coastal section, 370 m east-north-east of old kiln at Catcraig [7196 7736], 0.56 m of brownish-grey, very fine-grained limestone with some large productoids and crinoid debris.

From 2.34 to 7.32 m of strata separate the Lower Skateraw Limestone from the Middle Skateraw Limestone. This interval is made up of virtually equal proportions of silty mudstone and root-beds. A thin coal at the top of this interval is underlain by root-beds, the lower part of which is of ganister. The coal seam, about 0.15 m thick, is overlain by a thin development of dark grey mudstone which is in turn overlain by the Middle Skateraw Limestone.

Middle Skateraw Limestone

The Middle Skateraw Limestone ranges from 3.21 to 7.62 m in thickness, averaging about 5.15 m. The limestone is generally grey and brownish-grey, and unevenly bedded. Irregular lenticular calcareous siltstone or mudstone partings separate the harder limestone bands. Characteristically this limestone yields a rich fauna of rugose corals, brachiopods, crinoid debris, and an especially diagnostic band of the foraminifer *Saccamminopsis fusulinaformis* about 0.25 m below the top. This limestone is extensively worked for cement manufacture at Dunbar Quarry. Good exposures are seen at the following localities, sections which also illustrate the consistent development of this limestone over much of the area:

a Foreshore from Torness Point to Skateraw Harbour [7450 7542], 4.4 m of grey, hard, compact, flaggy limestone, gently undulating, irregularly bedded with dark grey calcareous siltstone or mudstone partings. Fossils tend to be more common in the less pure limestone bands, and particularly at the top and bottom of the development. *Saccamminopsis fusulinaformis* occurs in a band near the top.
b Foreshore 400 m east of old kiln at Catcraig [7190 7728], 4.1 m of fine-grained, pale brown-grey limestone with dark grey, calcareous siltstone bands. *Saccamminopsis fusulinaformis* is prominent in a band near the top, with rugose corals, some *Lithostrotion* colonies, brachiopods and crinoid debris.
c Railway cutting [6990 7671], 4.88 m of limestone with irregular dark grey, calcareous, siltstone bands, undulating bedding and rather nodular layers, the top is not exposed.
d North of White Sands the Middle Skateraw Limestone is exposed north-east of The Vaults from high tide to low water mark [7046 7829]. Here some 0.62 m of grey and brown-grey, rather impure limestone is exposed, yielding rugose corals, spiriferids, crinoid columnals and on the upper surface abundant *cauda-galli* markings. The *Saccamminopsis fusulinaformis* band was not recorded, but the general fauna and the stratigraphical setting of the limestone indicates that it is the Middle Skateraw Limestone.

Overlying the Middle Skateraw Limestone are up to 15 m of mainly shaly mudstone and silty mudstone. These beds are calcareous and very fossiliferous, with impersistent limestone bands in the lower part. The beds become less calcareous and less fossiliferous upwards. One thin limestone band about 3 to 4 m above the top of the Middle Skateraw Limestone, was previously named the Upper Skateraw Limestone. As this limestone occurs within the same marine cycle as the Middle Skateraw Limestone it is considered that this impersistent development should not be named. Additionally, in places up to three thin impersistent lenticular limestones are present in the calcareous beds overlying the Middle Skateraw Limestone, none of which can be termed distinctive. This calcareous development overlying the Middle Skateraw Limestone is extremely fossiliferous yielding a rich fauna which has been correlated with the Neilson Shell Bed of the Glasgow area (Wilson, 1974, p. 45). The development is exposed at the following localities:

a Skateraw Harbour [7385 7550] just over 7 m of dark-grey mudstone, mostly calcareous, with up to three impersistent limestone bands up to 0.45 m thick, yielding a rich fauna of corals, brachiopods, bivalves and trilobite pygidia.
b Foreshore east of Catcraig [7193 7724], where just over 9 m of mainly silty, calcareous mudstones and siltstones with one thin 0.15 m limestone, containing a rich shelly fauna.

Numerous borehole records indicate that the strata detailed above are persistent, with minor variations, over the whole area.

Upwards the lithology persists as dark grey mudstone or silty mudstone, in parts calcareous, but generally increasingly non-calcareous upwards. Gradual upwards-coarsening of the sediments takes place through siltstone and silty sandstone to sandstone. These strata are interesting in that they are patchily carbonate-cemented and characterised by numerous *cauda-galli* markings, worm-tracks and burrows, with some parts yielding rhynchonelloid fragments. This fauna is probably indicative of mainly quasi-marine conditions with occasional short-lived marine incursions.

At the northern end of the basin at Fluke Dub [6938 7842] and The Vaults [7025 7827], one marine incursion has led to the formation of a thin sandy limestone about 0.6 m thick with crinoid debris, an extremely local and impersistent development. Towards the upper part of these sandy beds, planty and rooty sandstones become prevalent and the beds towards the top of this sequence are of sandy and ganister-like seatearth. A few of the coarser sandstones are erosive and cut out the more marine beds. This is especially well seen near Barns Ness Lighthouse [7210 7733]. These beds indicate a fluvial, fresh-water environment when plants became established, although no coal seams were developed.

Subsidence leading to a prolonged marine incursion and the deposition of the Chapel Point Limestone followed this period of shallow water deposition.

Chapel Point Limestone

The Chapel Point Limestone ranges up to about 2.9 m in thickness. Generally this limestone is very hard, compact, fine-grained and impure, the basal part being often sandy. The fauna from this limestone is poor, mainly crinoid debris, poorly preserved shell debris and *cauda-galli* markings. Northwards the limestone becomes increasingly sandy and less fossiliferous, crinoid debris being the only fossils recorded.

Exposures of the Chapel Point Limestone are seen at the following localities:

a Chapel Point [7397 7585], where 2.9 m of well-bedded, grey, hard, compact limestone is well exposed at the type locality. Abundant crinoid debris, shell fragments and abundant *cauda-galli* markings are present.
b Foreshore, north of Barns Ness Lighthouse [7217 7758], where 2.4 m of grey, fine-grained limestone with brachiopods, crinoid debris and abundant *caudi-galli* markings, form prominent ledges to low water mark.
c White Sands [7096 7744, 7087 7805 and 7090 7808], a series of exposures, repeated by faulting, of limestone up to 2.4 m thick, grey and brown-grey, with crinoid debris, brachiopods, some small rugose corals and abundant *cauda-galli* markings.
d Foreshore north of The Vaults [7025 7845], the most northerly exposure seen; sandy limestone forms a ledge down to low water mark, crinoid debris and *cauda-galli* markings being the only organic traces noted.

Strata overlying the Chapel Point Limestone are mainly sandstones, in places markedly cross-bedded, often with ironstone nodules and patchy sideritic cement. A few thin, grey, siltstone bands, also with ironstone nodules, occur. These arenaceous beds are further characterised by roots and plant debris. The sandstone near the top of this interval becomes very hard, siliceous and ganister-like, a persistent feature of the sequence. The above interval between the Chapel Point Limestone and the Barns Ness Limestone varies from about 9 to 11 m in thickness.

Barns Ness Limestone

The Barns Ness Limestone, formerly the Barness East Limestone (Clough and others, 1910, p. 135), is grey and brownish-grey, fine-grained, hard and compact, somewhat dolomitic and relatively unfossiliferous. Crinoid debris is fairly common and *cauda-galli* markings well developed on weathered bedding planes. The limestone varies in thickness from 0.3 to 1.8 m, and usually forms a well-developed ledge, particularly from Barns Ness Lighthouse south-east to just north of Chapel Point. North of the lighthouse the limestone thins out rapidly, and dies out completely towards low water mark. To the north of White Sands, up to 0.9 m of limestone occurs [7067 7814], folded in a gentle syncline. *Cauda-galli* markings and crinoid debris are the only fossils present, but from lithological and stratigraphical evidence this is thought to be the Barns Ness Limestone. The sequence is repeated northwards, but here the limestone is absent, and the horizon is probably represented by a cannely, dark grey shale with *Lingula* resting on 0.61 m of ganister-like sandstone.

Strata above the Barns Ness Limestone are exposed to the north and east of Barns Ness Lighthouse, and north-east of Fluke Dub [6946 7828]. At the latter locality the strata are for the most part poorly exposed, the few exposures seen being of massive, cross-bedded sandstone with no particular characteristics. East of Barns Ness Lighthouse up to about 30 m of medium grained pale red-brown sandstone with prominent barrel-shaped sideritic 'doggers', forms a wave-cut platform down to low water mark.

Elsewhere in central Scotland, the top of the Lower Limestone Group is drawn at the Top Hosie Limestone or its equivalent, an horizon usually easily recognised by its lithological and faunal features. The Barns Ness Limestone may be the correlative of the Top Hosie Limestone but it may equate with one of the lower Hosie limestones. Supporting evidence on lithological or faunal grounds for either view is lacking. If the Barns Ness Limestone is equivalent to the Top Hosie Limestone then the strata above the limestone mentioned above should be placed in the limestone Coal Group of Namurian age. On the present evidence, however, there is no proof that Namurian rocks are present in the district.   AD

CHAPTER 6

Carboniferous palaeontology

INTRODUCTION

The fossiliferous nature of the Lower Carboniferous rocks on the coast south-east of Dunbar has long been known. Lists of the fossils from these coastal exposures and from the old quarries in the district were given by Salter (*in* Howell and others, 1866, pp. 74–77) and by Lee (*in* Clough and others, 1910, pp. 206–217) in the accounts related to the first two surveys of the district. In addition, Crampton (1905) referred to the characteristic species present in the limestones at Dunbar when comparing the sequences of the Lothians with those of Fife.

These works dealt mainly with the fossils in the uppermost part of the succession, the beds of which are well exposed at the surface. An important paper by H. H. Wilson (1952) described the fauna of the Cove Marine Bands found in the Thornton Burn. In the last decade, a series of IGS boreholes provided the opportunity to examine the fossil content of the whole of the Lower Carboniferous sequence in the area. The faunas from these borings and from the surface exposures were included in a paper by R. B. Wilson (1974) and these are treated in a general manner only in this present account.

CLASSIFICATION AND ZONATION

The strata dealt with in this section are those assigned to the Devono-Carboniferous, the Calciferous Sandstone Measures and the Lower Limestone Group. This classification is based on lithological criteria. Wilson (1974, p. 38) proposed that the Calciferous Sandstone Measures be divided into Lower and Upper Lothian groups, the boundary being drawn at the Macgregor Marine Bands which mark the earliest major marine incursion in the area. This proposal is not used in the present classification, however, as the division proved to be impractical for mapping purposes.

As the sedimentary succession is composed of cyclothems, the rock types are constantly changing when traced vertically. Also, there is a gradual shift from a non-marine to a marine environment through Dinantian times in the area as is evidenced by the fossils present. Consequently there is no group of animal fossils present throughout the succession which can be used for zonation purposes.

The rich marine faunas of the Lower Limestone Group, at the top of the succession, denote a P_2 age in the goniatite zonation (Currie, 1954, p. 534) and a Brigantian age in the classification of George and other (1976, p. 47). Below this there is little evidence from marine fossils to aid zonation. The only significant record is that of a goniatite from the Cove Lower Marine Band (Wilson, 1952, p. 312) to which Currie (1954, p.530) assigned a B age. The marine band is a component member of the Macgregor Marine Bands. The horizon is placed in the Asbian of the classification of George and others (1976, p.47).

The only fossils present throughout most of the local Carboniferous sediments are the numerous species of miospores and these were studied by Neves and others (1973). In that work, a classification of the Lower Carboniferous of Scotland and northern England was proposed, based on the palynological study of the succession in the IGS borehole at Spilmersford, to the west of the present area (Neves and Ioannides, 1974). The succession was divided into five miospore assemblage zones, but two miospore zones known to occur at the base of the Carboniferous in southern England were not recognised. This suggests that the two missing zones may be represented by part at least of the conformably underlying Old Red Sandstone facies rocks, here called Devono-Carboniferous. It is unfortunate that these sediments have so far failed to yield spores.

FAUNAS

Devono-Carboniferous

The sediments here placed in the Devono-Carboniferous have yielded fossils from only two localities in the area. The first is that cited in Clough and others (1910, p. 35) where it was stated that a scale of the fish *Holoptychius nobilissimus* was found in a fallen block in a gorge near Whittingehame. The other locality is the IGS borehole at Birnieknowes where a similar scale occurred at c.436 m and indeterminate fish scales were found at a slightly lower horizon.

Calciferous Sandstone Measures

The fossils present in the Calciferous Sandstone Measures show that a considerable thickness of sediments was deposited in non-marine environments before the first major marine transgression affected the area. This sequence was well developed in the IGS Birnieknowes Borehole, near the eastern margin of the area, where c.350 m of strata occurred between the top of the red sandstone facies and the lowest major marine band. In the IGS East Linton Borehole, in the western part of the area, some 380 m of similar rocks were proved and the red sandstone facies was not reached. The fossils identified from these sequences of the lower part of the Calciferous Sandstone Measures comprise plant fragments, including *Alcicornopteris sp.*, megaspores, *Spirorbis sp.*, the gastropod *Naticopsis*? *scotoburdigalensis* (Etheridge jun.), the bivalves *Modiolus* cf. *latus* (Portlock) and *Naiadites obesus* (Etheridge jun), estheriids, ostracods and fish remains. In the Birnieknowes Borehole, two poorly developed marine bands, containing little more than *Lingula*, were found in the upper part of these beds. This suggests that marine conditions existed nearby but could not become established in the area.

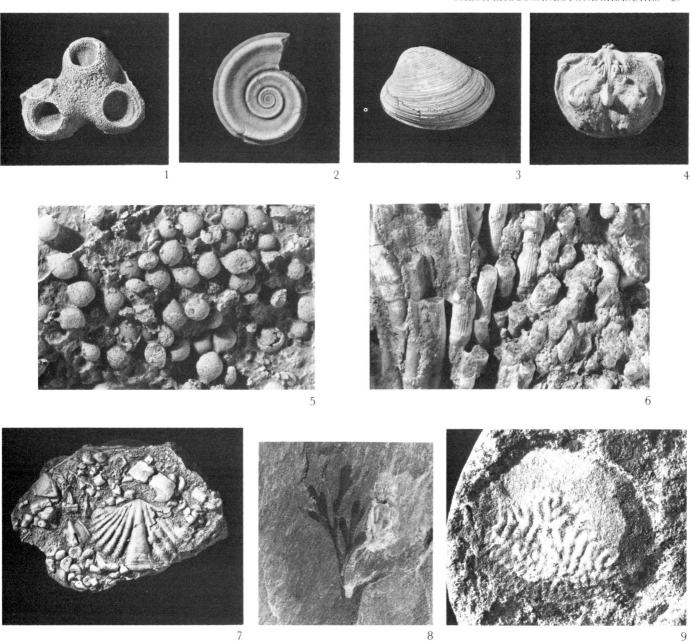

Plate 8 Devono-Carboniferous and Carboniferous fossils

1 *Microcyathus cyclostomus* × 2, Middle Skateraw Limestone, Skateraw Quarry.

2 *Straparollus (Euomphalus) carbonarius* × 2, Middle Skateraw Limestone, East Barns Quarry.

3 *Nuculopsis gibbosa* × 2, Middle Skateraw Limestone, East Barns Quarry.

4 *Eomarginifera setosa* (brachial valve) × 2, Middle Skateraw Limestone, East Barns Quarry.

5 *Saccamminopsis fusulinaformis* × 3, Middle Skateraw Limestone, shore at Catcraig.

6 *Lithostrotion junceum* × 2, Upper Longcraig Limestone, Catcraig Sea Quarry.

7 *Punctospirifer* cf. *scabricosta* × 2, Macgregor Marine Bands (Linkhead Limestone), shore at Linkhead.

8 *Telangium affine* × 1, Macgregor Marine Bands, Birnieknowes Borehole at 57.28 m.

9 *Holoptychius nobilissimus* (scale) × 1, Devono-Carboniferous, Birnieknowes Borehole at 463.60 m.

The earliest major Carboniferous marine transgression in central Scotland is marked by the occurrence of the Macgregor Marine Bands (Wilson, 1974, pp. 41–43). These are a group of marine bands, up to three in number, found in the east of Scotland as far north as Fife and they contain several distinctive species. In the present area representatives of the bands occur on the coast at Linkhead, the Linkhead Limestone, at Thornton Burn, probably in two bands (Wilson, 1952; Wilson, 1974), in the IGS Birnieknowes Borehole at c.45 m and in the IGS Skateraw Borehole at 150 and 167 m. The combined faunas from these localities include the corals *Heterophyllia sp.* and *Lithostrotion junceum* (Fleming), the brachiopods *Buxtonia sp.*, *Gigantoproductus sp.*, *Isogramma sp.*, and *Punctospirifer* cf. *scabricosta* North, the bivalves *Cypricardella sp.*, *Edmondia sulcata* (Fleming), *Leiopteria hendersoni* (Etheridge jun.), *Palaeoneilo mansoni* Wilson, *Sanguinolites clavatus* (Etheridge jun.), *Streblopteria? redesdalensis* (Hind), and *Wilkingia elliptica* (Phillips), the goniatite *Beyrichoceratoides redesdalensis* (Hind) and the echinoid *Archaeocidaris sp.*

The Lower and Middle Longcraig limestones at the top of the Calciferous Sandstone Measures mark the inception of the dominant marine facies which continued through the overlying Lower Limestone Group. During this period of deposition, marine influences were at their maximum in Scotland and rich marine assemblages are present in the limestones and associated mudstones. These beds are exposed at several localities in the area and can be examined at the coastal sections especially at White Sands and Skateraw.

The Lower Longcraig Limestone contains a rich brachiopod fauna with the maximum development in the succession of the large inrolled productoid *Semiplanus* cf. *latissimus* (J. Sowerby). The Middle Longcraig Limestone is rich in corals with colonies of *Lithostrotion spp.* and large solitary forms such as *Dibunophyllum*. It also contains numerous specimens of the brachiopods *Composita* and *Pleuropugnoides*. The upper layers of the limestones are altered to a light grey or greenish mud and the fossils are obscured by recrystallisation. This phenomenon is best explained by the presence of the seatearth of a thin coal immediately overlying the limestone and stigmarian roots from the seatearth have been observed to be impressed on the top surface of the limestone. It seems probable that the leaching effect of plants growing in a soil overlying the limestone effected an alteration in it, perhaps before its complete consolidation.

Lower Limestone Group

The Upper Longcraig Limestone, at the base of the Lower Limestone Group, contains a rich and distinctive fauna. The mudstones immediately underlying it are characterised by large specimens of the brachiopod *Orbiculoidea* and by a bivalve assemblage including *Actinopteria persulcata* (McCoy), *Naiadites crassus* (Fleming) and *Streblopteria ornata* (Etheridge jun.). The limestone itself carries a rich fauna, dominated by brachiopods and corals.

The next limestone in ascending order is the Lower Skateraw which is a relatively thin bed and poorly fossiliferous. It is characterised, however, by *Gigantoproductus* cf. *giganteus* (J. Sowerby) which occurs at almost every exposure where the limestone has been examined. The succeeding marine phase is represented by the Middle Skateraw Limestone and the associated calcareous mudstones. This limestone was quarried extensively in the area and over the last century a very rich and varied fauna has been obtained from it including some very rare species. The limestone itself is noted for a band near the top of the bed, which is largely composed of the foraminifer *Saccamminopsis fusulinaformis* (McCoy). Numerous species occur in the calcareous mudstones which form partings in the limestone and lie above the bed. These include the rare coral *Microcyathus cyclostomus* (Phillips), the brachiopods *Crurithyris urii* (Fleming), *Eomarginifera setosa* (Phillips) and *Tornquistia youngi* Wilson, the gastropods *Glabrocingulum atomarium* (Phillips), *Straparollus* (*Euomphalus*) *carbonarius* (J. de C. Sowerby) and *Tropidocyclus oldhami* (Portlock) and the bivalve *Euchondria neilsoni* Wilson. Some of these species are diagnostic of the Neilson Shell Bed (Wilson, 1966), a marine band which occurs in Fife and in most parts of the Scottish Central Coalfield, and their presence in the Middle Skateraw Limestone leads to its correlation with a horizon found over much of central Scotland.

The upper part of the Lower Limestone Group in the area is unusual in its development when compared with contemporary rocks in other parts of Scotland. Elsewhere it contains numerous rich marine bands but in the Dunbar area only two marine limestones are present. These are the Chapel Point and Barns Ness limestones which are poorly fossiliferous, the few specimens observed being mainly large bellerophontid gastropods and coiled nautiloids.   RBW

CHAPTER 7

Garleton Hills Volcanic Rocks

INTRODUCTION

Only the lower part of the Garleton Hills Volcanic Rocks, which lie within the Calciferous Sandstone Measures, are present in the district. The full sequence crops out to the west in the adjoining Haddington district (McAdam and Tulloch, 1985). The volcanic rocks occur in the north-western part of the district and overlie the cementstone facies sediments which form the lowest part of the Calciferous Sandstone Measures. The Garleton Hills Volcanic Rocks are approximately contemporaneous with the Arthur's Seat Volcanic Rocks of the Edinburgh District (Mitchell and Mykura, 1962, pp. 45–52). Relationships with the volcanic rocks across the Lammermuir Fault at Oldhamstocks, near the eastern margin of the district and at Fluke Dub, near Dunbar, are uncertain.

OUTCROP

Gentle folding along north-east to south-west axes has divided the outcrop of the Garleton Hills Volcanic Rocks into three separate areas. In the north in the Whitekirk Syncline, the northern limb of which is cut out by the Gleghornie Fault, the basalt lavas are only exposed in the Whitekirk Hill–Bankhead area. The underlying tuffs crop out on the shore at The Gegan, north of this fault. Further south, between the Crauchie and Traprain anticlines, lies a complex basin around East Linton with basaltic tuffs and lavas outcropping in the Lawhead–East Linton area and east of Kippielaw. On the south side of the Traprain Anticline, basaltic and trachytic rocks dip southwards against the Dunbar–Gifford Fault. Numerous agglomerate-filled necks are exposed along the coast, particularly north-west of Peffer Sands and around Dunbar. Volcanic plugs are rare, but intrusions of many forms, contemporaneous with the volcanics or younger, are common (Figure 10).

STRATIGRAPHY

In the Haddington district, subdivision of the Garleton Hills Volcanic Rocks into members has been carried out on lithological grounds (McAdam and Tulloch, 1985). The members (Table 1) persist laterally into the present district, though they may interdigitate or have diachronous boundaries.

The basal basaltic tuffs of the North Berwick Member are thickest in the north where they can be divided into lower green and upper red subdivisions. The East Linton Member is characterised by augite-phyric olivine-basalts of Craiglockhart type (ankaramites) and Dunsapie type, with trachybasalts such as kulaites and mugearites. The Hailes Member is composed dominantly of feldspar-phyric olivine-basalts of Markle type (hawaiites) with mugearites. Locally

in the south, trachyte lavas of the Bangley Member are preserved.

Table 1 Classification of the Garleton Hills Volcanic Rocks

BANGLEY MEMBER	Trachyte lavas (0–20 m)
HAILES MEMBER	Markle basalts and mugearites (40 m)
EAST LINTON MEMBER	Mostly Dunsapie and Craiglockhart basalts, mugearites and kulaites (20–90 m)
NORTH BERWICK MEMBER	Red basaltic tuffs, green basaltic tuffs, freshwater limestones and cementstones (20–220 m)

EARLY VOLCANICITY

Precursors of the Garleton Hills volcanic activity were recorded in the underlying cementstone facies in the IGS East Linton Borehole. There, pyroclastic beds were found to be interbedded with sediments over some 77 m of strata. Equivalent activity was represented by only 5 m of tuffaceous beds at Spilmersford some 15 km to the south-west in the Haddington district. The pyroclastic beds recorded in the East Linton Borehole, over 100 m below the base of the North Berwick Member, were as follows:

	Thickness m
Purple, tuffaceous siltstone	0.7
Strata	4.5
Purple and grey-purple, bedded, tuffaceous siltstone and sandstone	5.2
Strata	33.6
Grey, bedded, tuffaceous siltstone and tuff	5.5
Strata	7.2
Grey-purple, bedded tuff	1.4
Strata	11.3
Green-grey, poorly bedded, coarse tuff with lava fragments up to 10 cm; basal agglomerate containing lava, cementstone and siltstone fragments	7.9

NORTH BERWICK MEMBER

Pyroclastic rocks of the North Berwick Member form the basal beds of the Garleton Hills Volcanic Rocks from the coast to the Dunbar–Gifford Fault. The thickest development, around 200 m, is in the north where the rocks crop out round the Whitekirk Syncline. Lower green beds are recorded only in the north where they are excellently exposed in coastal cliffs around The Gegan. Inland exposures are poor along the irregular outcrop round the

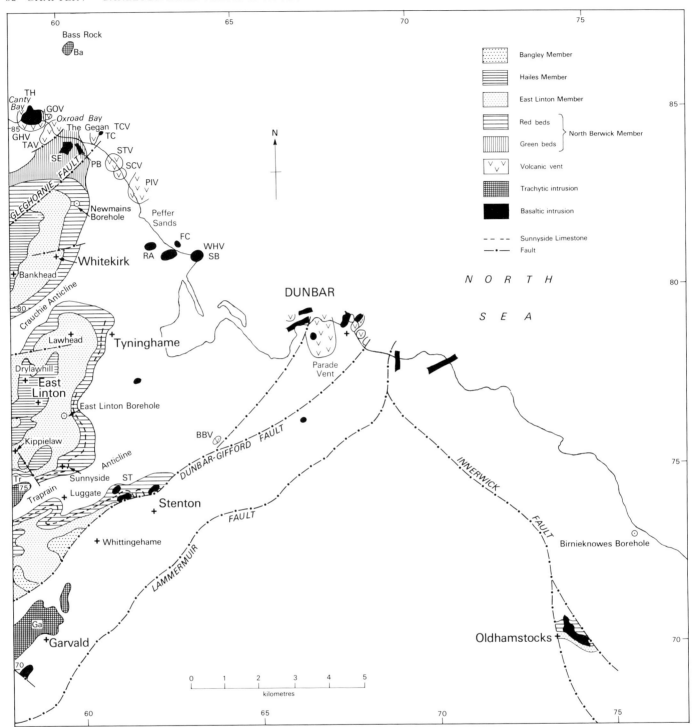

Figure 10 Generalised map of the Garleton Hills Volcanic Rocks, vents and intrusions in the district

Abbreviations Ba - Bass Rock Plug; BBV - Belton Brae Vent; FC - Frances Craig Sill; Ga - Garvald Sill; GHV - Gin Head Vent; GOV - Gin Head (Older) Vent; PB - Primrose Bank Sill; PIV - Pillmour Vent; RA - Ravensheugh Sill; SB - St Baldred's Plug; SCV - Scoughall Vent; SE - Seacliff Plug; ST - Stenton Sill; STV - Seacliff Tower Vent; TAV - Tantallon Vent; TC - The Car Dykes; TCV - The Car Vent; TH - Taking Head Sill; Tr - Traprain Law Laccolith; WHV - Whitberry Vent

Crauchie and Traprain anticlines and the Prestonkirk and Kippielaw synclines. As it is traced to the south, the member becomes thinner, all of the beds are red and it contains the freshwater Sunnyside Limestone, as shown in the section through the member in the East Linton Borehole (Figure 11). Numerous tuff- and agglomerate-filled necks and intrusive plugs were formed during this period, particularly around Dunbar and on the coast between Peffer Sands and Canty Bay.

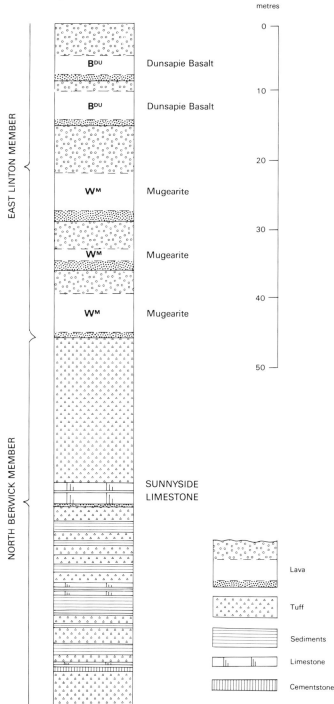

Figure 11   North Berwick Member and part of East Linton Member, IGS East Linton Borehole

DETAILS

Oxroad Bay to The Gegan

Cliffs up to 15 m high give excellent exposures of the basal green tuffs in this area (Martin, 1955; Kelling *in* Mitchell and others, 1960). The green, fine- to coarse-grained tuffs are well bedded and water-lain with impersistent bands of ripple-bedded sandstones and cementstones containing small carbonised plant fragments and rare bivalve casts. There are also fragments of ultrabasic rocks.

In the vicinity of the Hillhouse basalt Seacliff Plug the tuffs are affected by staining. A sharp, irregular junction between red and green tuffs cuts across the bedding on the cliffs on both sides of the point opposite The Gegan and can be traced across the foreshore. This is interpreted as the original green tuffs having been stained red. At The Gegan stack (Plate 9) and the cliffs opposite there are red-brown stained, bedded, coarse-grained agglomeratic tuffs with bombs of porphyritic and non-porphyritic basalt and mugearite. Blocks of marl and limestone are also present. These are quite different from the other green tuffs as they lack cementy and sandy ribs. The tuffs at The Gegan are cut by fault-joints, with calcite veining and sedimentary dykes.

Whitekirk

There are no outcrops of red tuff round the Whitekirk Syncline, but red and purple soils show that red tuffs are near the surface. The Newmains water borehole, starting within 10 m of the base of the lavas, proved some 70 m of red, calcareous, tuffaceous mudstones with coarse-grained tuff bands.

East Linton

In the East Linton area there is little outcrop evidence but a complete section was proved in the East Linton Borehole (Figure 11), including the Sunnyside Limestone, 2.9 m thick. Below the limestone there are tuffaceous red-stained calcareous sediments 55.4 m thick. The beds, variously red, red-purple, grey-purple and green-purple in colour, include mudstones, silty mudstones, siltstones, thin cementstone and limestone ribs and nodules. The bedding has been destroyed in places by the reddening which has coloured the cementstones dark red, and produced reduction spots and pseudobrecciation. Above the limestone the entire 21.7 m of beds are tuffs which are grey-purple, red-stained, calcareous, fine- to coarse-grained, and bedded in places. Thin cementstone bands and nodules occur and the tuffs are agglomeratic, particularly near the top, with vesicular lava fragments up to 5 cm. Thin calcite veins are common. Red, bedded tuffs, 2 to 3 m thick, were recorded above the limestone in infilled quarries 300 and 600 m south of the railway.

The type locality of the Sunnyside Limestone is a series of quarries round Sunnyside Farm (Figure 10), now mainly obscured or infilled. Large quarries between Sunnyside and East Linton have also been filled in. About a metre of limestone is exposed in the railway cutting east of East Linton. In the East Linton IGS Borehole (Figure 11) the detailed section showed that the limestone occurred in two leaves:

	Thickness m
Tuff, red	
Limestone, pale purple and purple, banded with some cross-bedding, very hard, porcellanous	1.14
Mudstone, tuffaceous, purple	0.20
Limestone, purple, unbedded, hard, tuffaceous	1.55
Mudstone, grey-purple, silty	

Plate 9   Bedded agglomeratic tuff of the North Berwick Member, showing prominent infilled joints, The Gegan (D 1130)

The limestone outcrop can be traced west of Tyninghame along a north–south line of old quarries. Further north at North Berwick, in the adjacent Haddington district, the Rhodes Limestone and the calcareous horizon between the green tuffs and red tuffs of the North Berwick Member may be equivalent to the Sunnyside Limestone.

On the south side of the Traprain Anticline the Sunnyside Limestone has been mapped along the outcrop from the edge of the district west of Luggate to the Dunbar–Gifford Fault. Several old quarries exploited the limestone. Typically it is a freshwater, banded limestone, stained red, pink and white. No fossils have been found in the bed.

Whittingehame – Stenton

In the Whittingehame area the tuffs are not exposed and are probably thin. Near Stenton the tuffs thicken and are seen in sections in the Sauchet Burn where at least 7.5 m of purple, bedded, medium- to coarse-grained tuff is associated with purple sandstone. Sections in tuffs can also be seen in the valleys of the Whittingehame Water and Luggate Burn near their confluence.

EAST LINTON MEMBER

The lowest flows of this member are usually non-porphyritic trachybasalt, either mugearite or leucite-kulaite. These are succeeded by five or six flows of macroporphyritic olivine-basalt of either Craiglockhart type or Dunsapie type. In the district these lavas form three areas of high ground where they are well exposed, namely Whitekirk Hill, from Lawhead to Sunnyside, and a small area east of Whittingehame. The basaltic flows are commonly 5 to 15 m thick with as much as half of each flow comprising a soft, slaggy, vesicular, upper part. Typical trap-featuring is normally well displayed, being produced by selective weathering of the hard and soft parts of the flows.

DETAILS

Whitekirk Hill

At least five flows of the East Linton Member, resting on red basaltic tuffs, form the stepped, eastern face of Whitekirk Hill, where they are cut by a small dip fault. The lowest flow forms a low scarp, is probably only a few metres thick, and is apparently overlapped by the succeeding flow. It is an unusual trachybasalt, almost destroyed by weathering, which can be compared with the kulaite of North Berwick and Blaikie Heugh (McAdam and Tulloch, 1985). The second flow, probably 9 to 12 m thick and forming a prominent continuous scarp, is a fairly fresh olivine-basalt of Craiglockhart type. Above this are scarps of at least three flows of fairly fresh olivine-basalt of Dunsapie type, the topmost forming the dip slope on the east of Whitekirk Hill.

Lawhead

North of East Linton there are good exposures of the lower flows of the member in the north and east faces of Lawhead Hill. Elsewhere in the area, the lavas are either drift-covered or occur in isolated knolls. The lowest flow is seen in a small quarry [604 796] north of The Mast where 3.7 m of red-purple, mottled, very rotten trachybasalt lava is exposed. This is perhaps equivalent to the lowest flow on Whitekirk Hill. The second flow is mugearite, seen in 7-m high crags along the north face of Lawhead Hill and in quarries at The Mast and north-west of Lawhead. It is a dark, purple-grey, non-porphyritic mugearite. Forming a higher scarp on the north-west slope of Lawhead Hill and also the high ground around the farm is a flow, at least 3 m thick, of dark purple-grey basalt of Craiglockhart type. The succeeding Dunsapie basalt lavas do not form scarps but occur as knolls protruding through the till.

East Linton

The IGS East Linton Borehole penetrated the lowest five lavas of the East Linton Member (Figure 11). Three mugearite lavas, respectively 10, 8 and 14 m thick were followed by two basalt lavas of Dunsapie type, each almost 8 m thick. The lavas dip at $15° - 20°$ and show a great variation in the proportion of the massive central part to the top and basal amygdaloidal parts. The succession recorded is as follows:

	Thickness m
Flow 5 Basalt of Dunsapie type: purple with calcite, chloritic, limonitic and opaline amygdales in top 4.45 m; grey-purple, with large and small plagioclase, augite and olivine phenocrysts in centre; purple slaggy with calcite and chloritic veins in basal 0.9 m	7.91 +
Tuff, purple	0.01
Flow 4 Basalt of Dunsapie type: grey-purple with calcite and limonitic amygdales, vugs lined by dog-tooth spar, chloritic veins in top 1.6 m; grey-purple and red-purple, massive, with large and small plagioclase, augite and olivine phenocrysts in centre; grey-purple with calcite amygdales and veins in basal 0.7 m	7.63
Flow 3 Mugearite: red-purple and grey-purple, with calcite and limonitic, mainly elongate amygdales in top 7.15 m; red-purple and purple, fine-grained, non-porphyritic in centre; grey-purple with calcite amygdales in basal 1.2 m	14.00
Flow 2 Mugearite: dark purple with calcite amygdales and veins in top 4.35 m; dark purple and grey-purple, fine-grained, non-porphyritic in centre; purple with calcite amygdales in basal 1.35 m	8.11
Flow 1 Mugearite: purple and grey-purple with calcite amygdales in top 3.75 m; dark-purple, grey-purple and brown-purple, fine-grained, non-porphyritic in centre; grey-purple, red-purple and yellow with calcite amygdales in basal 0.55 m	10.06
Tuff, purple-grey	

In the adjacent railway cutting two flows of mugearite lava are exposed, and at least four flows of olivine-basalt overlie the mugearites. The lowest is of Craiglockhart type, the next two are of

Dunsapie type and the highest flow is intermediate between Dunsapie and Markle types.

The narrow steep-sided gorge, 20 to 25 m deep, cut by the River Tyne south-west of East Linton exposes two higher flows of the East Linton Member lying below the basal flows of the Hailes Member. The lower flow exposed in the river bed is possibly equivalent to Flow 4 in the East Linton Borehole and is a dark basalt of Dunsapie type with large feldspar and augite phenocrysts and small, pseudomorphed, olivine phenocrysts. The higher flow forms most of the sides of the gorge and is present in areas to the north around Drylawhill and the slope up towards Kippielaw. It is at least 10 m thick, and is a more feldspathic Dunsapie basalt, with small feldspar phenocrysts and a few augite and olivine phenocrysts.

Sunnyside

South from East Linton the flows become thinner. Mugearites and Craiglockhart and Dunsapie type basalts can still be identified but as the scarps are cut by small faults and glaciated, the relationships between isolated outcrops are not clear. West of Sunnyside one flow of mugearite is present at the base, succeeded by two basalts, the lower being of Dunsapie type and the other intermediate between Markle and Dunsapie types.

Whittingehame

In this area away from the volcanic centre, the lavas tend to die out and only two flows are definitely present, a leucite-kulaite succeeded by a Craiglockhart type basalt. The former is an unusual purple trachybasalt, well exposed in Blaikie Heugh.

HAILES MEMBER

The flows of this member, about six in number, are feldsparphyric basalts of Markle type (including the type Markle locality). Non-porphyritic to sparsely porphyritic basalts and mugearites are also present. In the district the member occurs in synclinal cores in three isolated areas. These are west of Whitekirk Hill, in the high ground around Drylawhill and east of Kippielaw. Although these rocks form some good scarps with trap-featuring, glaciation has commonly reduced them to complex patterns of isolated knolls.

DETAILS

Bankhead

South of the Gleghornie Fault there are several flows of the Hailes Member dipping to the west-north-west on the south-east limb of the Whitekirk Syncline. These comprise three or four feldsparphyric flows of olivine-basalt which are overlain by a flow of mugearite 10 m thick. Some of the basalts are of characteristic Markle type. The mugearite flow forms only isolated knolls.

Drylawhill to East Linton

The area around East Linton has a particularly complex outcrop pattern of basaltic flows. The succession is fairly simple with two distinctive basalts more or less related to the Markle type. The lower flow (or flows) in the type Markle basalt (MacGregor, 1948), including that in Markle Quarry itself, consists of 6 m of basalt with very abundant feldspar phenocrysts. The outcrop extends along the north limb of the syncline from Drylawhill to the slopes between the A1 road and the River Tyne. In the latter locality a flat-lying higher flow, with sparse feldspar phenocrysts, caps the top of the high ground. This flow also occurs at the top of Markle Quarry in a hollow eroded in the type Markle flow.

Kippielaw

A similar succession can be identified on the south side of the River Tyne on the south limb of the syncline. The same two flows, as at East Linton, are present and in addition two higher ones cap the hill between Kippielaw and Traprain. The type Markle basalt occurs along the north of the outlier, and may thin out completely along the south side. The basalt with sparse feldspar phenocrysts occurs but is poorly exposed round the outlier. A flow of mugearite caps the hill and forms a 3-m high scarp along the south side.

BANGLEY MEMBER

Two areas of trachyte lava occur west of Whittingehame, on the north side of the Dunbar–Gifford Fault. The trachyte rests directly on lavas of the East Linton Member overstepping the upper basalts of the Hailes Member. Exposures in the trachytes are poor and little detail is known of them. ADM

OUTCROPS SOUTH-EAST OF THE LAMMERMUIR FAULT

Two small areas of volcanic rocks, with associated intrusions, occur to the south-east of the Lammermuir Fault; one at Oldhamstocks and the other at Fluke Dub on the coast 1.5 km east of Dunbar.

The apparent relationships of the volcanic rocks at Oldhamstocks to the surrounding Devono-Carboniferous sediments, suggests that the age of the volcanics may be older than suggested by Clough (in Clough and others, 1910, p. 86), where the volcanics were tentatively placed at least some 150 m above the local base of the Carboniferous. These rocks are now thought to be considerably older than the Garleton Hills Volcanic Rocks.

The age of the tuffs, Dunsapie basalt and nepheline-basanite at Fluke Dub can only be determined from the assumed age of the neighbouring strata, and on this basis, it is probable that these igneous rocks are contemporaneous with the Garleton Hills Volcanic Rocks.

DETAILS

Oldhamstocks

In Wally Cleugh [7440 6980] a number of poor exposures of basalts, with rotten bands suggesting the presence of more than one flow, overlie red and grey-white friable sandstones referred to the Devono-Carboniferous. These basalts are apparently unconformably overlain by a sill-like intrusion of basanite, which is in turn succeeded downstream by a green, buff and red, tuffaceous sandstone, abruptly faulted downstream against Devono-Carboniferous sandstone [7447 6998]. In the Oldhamstocks Burn there is a series of exposures of purple, blue and green mottled red tuffs with sandstone ribs and fragments up to 0.1 m [7421 7033 to 7400 7045]. Some large blocks of friable sandstone also occur in the tuffs. Upstream, these tuffaceous beds are apparently overlain by strata of Devono-Carboniferous affinities. Downstream the tuffaceous beds are faulted against brown and red-brown sandstones with irregular, nodular, irony layers (poorly developed cornstones) of Devono-Carboniferous age. Both tuff and basalt lavas appear to be overstepped upstream by the basanite intrusion, although the lack of good exposure precludes the precise determination of the relationship.

Fluke Dub

On the coast 1.5 km east of Dunbar, a faulted area of sediments and volcanic rocks is very poorly exposed. Soft, rotten, brownish-red, fine-grained tuff, with much green, chloritic material, occurs in a broad, flat, weathered tract on the foreshore. A fault, trending almost north–south, downthrows strata of Lower Limestone Group age against the tuff to the east. To the west of the tuff, two vertical or near vertical basalt outcrops form a prominent north–south ridge. The more easterly one is of bluish-grey, rotten, chloritic olivine-basalt of Dunsapie type and is probably a lava-flow. Westwards this basalt is succeeded by a thin bed of red mudstone with green bands and few lenticular grey and red-stained cementstone ribs. Further to the west, the second outcrop is of rotten, soft basalt with irregular joints and thin veins of carbonate. It has been classified as a nepheline-basanite and the rock-type suggests that it is an intrusion. Sandstones, shales and cementstones of Calciferous Sandstone Measures age overlie the basanite to the west.

The fault to the east of the tuff at Fluke Dub extends to the south and cuts strata exposed in the railway cutting south of Broxmouth [6973 7676]. To the west of this fault are poor exposures of purple and red-purple tuff, up to about 25 m thick, flanked on both east and west sides by strata referred to the Calciferous Sandstone Measures. These strata, all vertical or subvertical, are considered to be part of the development seen at Fluke Dub. AD

VOLCANIC VENTS

Along the coast from Canty Bay to Peffer Sands there are no less than six large tuff- and agglomerate-filled volcanic vents. These were first recognised by Day (1928; 1930) who named them the Gin Head, Tantallon, The Car, Seacliff Tower and Scoughall vents and the Pillmour Volcano. In addition, Day's Whitberry Vent at the south end of Peffer Sands is almost filled by the St Baldred's Plug.

At Dunbar several vents exposed along the coast were described by Maufe (in Clough and others, 1910) and by Francis (1962; in Craig and Duff, 1975). By far the largest was named the Parade Vent by Francis and smaller ones are called the Belhaven Point, Coastguard Station, Kirk Hill and Old Harbour vents.

Martin (1955) divided the northerly vents into a younger Green Group including the Gin Head and Tantallon vents and an older Red Group containing the other vents. Evidence for their relative age occurs at Gin Head where a small Red Group vent is cut by the main Gin Head Vent of the Green Group. In the Haddington district to the west, Green Group vents cut strata which contain fragments of lavas up to the Hailes Member. Martin considered the Red Group vents were active in the period prior to the eruption of the lavas, as they only cut strata below the Garleton Hills Volcanic Rocks.

DETAILS (letters refer to location map (Figure 10)

Vents of the Red Group

Gin Head (older) Vent (GOV)

On the foreshore south-east of Gin Head this vent occupies a small fairly well-defined area cutting the Canty Bay Sandstone, and is itself truncated to the west by the later Green Group vent (q.v.). The vent is filled with soft, red and red-green, unbedded agglomerate containing abundant blocks of red tuff, and Day (1928) recorded cornstones, bombs of decomposed olivine-basalt and fragments of fossil wood from it.

The Car Vent (TCV)

Day (1930) interpreted this as a double vent forming the north part of The Car, whereas Martin (1955) recognised two further vents on the foreshore to the south. A small vent is also exposed near Low Water Mark to the east. At Car Beacon the northern orifice is filled with red-brown, agglomeratic tuff, poorly bedded with dips ranging from 60° to the NW up to vertical. The tuff is cut by several low-angle fault-joints, contains numerous highly vesicular basalt bombs and includes small basanite intrusions. The next vent at Great Car contains similar red-brown, agglomeratic tuff with highly vesicular basalt bombs as well as fragments of non-vesicular basalt and bedded tuff; it is poorly bedded, dipping to the north. Only the eastern margin is visible of the third vent which contains poorly bedded red, coarse- and fine-grained tuff. The southern vent near High Water Mark has red-brown, unbedded, agglomeratic tuff with small basalt bombs and blocks of sandstone from a few centimetres to several metres in size.

Seacliff Tower Vent (STV)

The eastern part of this vent is well exposed on the foreshore from Seacliff Tower to Chapel Brae (Day, 1930). The vent agglomerate forms a platform higher than, and sharply cutting, the bedded sediments. Normally the country rocks dip gently to west or northwest, but near the vent-margin the dips steepen markedly. Fine cliff sections show that the material infilling the vent is composed of coarse-grained, red, agglomeratic tuff containing blocks almost entirely of sedimentary rocks such as sandstone and marl. The tuff is poorly stratified with generally steep, inward dips, and is cut by prominent low angle joints.

Scoughall Vent (SCV)

On the foreshore just to the south of the last mentioned vent lies a similar but smaller one. Again, only its eastern part is exposed, sharply cutting bedded sandstones and marls. The vent contains red, unbedded, agglomeratic tuff with large bedded sandstone blocks and a few small fragments of decomposed basalt. Day (1930) recorded a dome-like structure produced by joints dipping gently away from the centre of the vent.

Pillmour Vent (PIV)

In the south part of Scoughall Rocks, Day (1930a) recognised a large ill-defined vent, its margin mainly obscured by beach deposits. The northern part contains red tuff with no bedding and few fragments. In the south there is dark red and red-brown, fine- to coarse-grained, bedded tuff, notable for blocks of sandstone, sandstone masses up to 100 m long, and several small basaltic intrusions.

Whitberry Vent (WHV)

Although virtually filled with a plug of Craiglockhart type basalt, in this vent there is some red agglomeratic tuff with vesicular basalt blocks along the southern margin. This suggests that the plug is filling an earlier Red Group vent (Day, 1930).

Belton Brae Vent (BBV)

In the valley of the Biel Water near Belton Brae, there are sections of up to 12 m of purple, coarse-grained, unbedded, agglomeratic tuff with numerous basalt bombs. No margins of this small vent are seen against the country rocks which are sandstones and mudstones.

Vents of the Green Group

Gin Head Vent (GHV)

The foreshore from Canty Bay to Gin Head provides a narrow section through the middle of this large vent (Day, 1928). Only small segments of its margin are seen at the west and east ends but in both places, though obscured, the vent-rocks strongly transgress the country rocks. Less than half the vent is occupied by green, coarse-grained, agglomeratic tuff, reddish towards the west, containing bombs and blocks of basalt, bedded tuff and sedimentary rocks. These features are best seen in the 15 m section at Gin Head. Much of the vent is occupied by the irregular Taking Head monchiquite intrusion (p. 45). The latter is mainly sill-like but some dykes are also present and their complex relationship with the agglomeratic tuff is well displayed on the foreshore. Other features of the vent are large masses of sandstone several tens of metres long, cementy ribs, and a fossiliferous limestone, 0.6 m thick, in the tuff near the east margin.

Tantallon Vent (TAV)

The foreshore and cliffs at Tantallon Castle provide an excellent section through the middle of the Tantallon Vent, first described by Day (1928). Its north-western margin clearly transgresses bedded sediments whereas the south-eastern margin is faulted against bedded green tuffs and sediments. The vent-material is predominantly of poorly bedded, green, coarse-grained, agglomeratic tuff containing numerous basalt bombs and blocks of bedded tuff. A few small basalt dykes and sills and fault-joints cut the agglomerate. Within the agglomerate below the north-western wall of the castle lies 6 m of pale green-brown, medium-grained, bedded sandstone, probably deposited while the vent was quiescent.

ADM

PETROGRAPHY

The nomenclature used on the geological map of the district and in this account is the petrographical classification of Carboniferous basalts formulated by MacGregor (1928). This classification divides the basaltic rocks into six main types, with transitions, depending mainly on the type and size of phenocrysts but also on the generally feldspathic or pyroxenic nature of the rock. A more recent classification (Macdonald, 1975) of such rocks was proposed, based on their chemical composition and hence on the normative minerals. Macdonald's correlation between the two classifications was given, with minor modifications, by Elliot (in McAdam and Tulloch, 1985, table 2).

The use of a classification based on the petrochemistry of the rocks is more helpful in a study of the geochemistry, provenance and petrogenesis of a volcanic suite. It is, however, less useful for mapping purposes unless large numbers of analyses are available and is not useful in dealing with altered rocks. MacGregor (1928, p. 348) pointed out that rocks similar, in many respects, to mugearites are produced by the more or less complete albitisation of some fine-grained basalts. Apart from the petrographic difficulties, such partial or complete albitisation can strongly affect a chemical classification and may cause an altered basalt to have the necessary composition to be classed as hawaiite or mugearite.

Basalts in the district are mainly of Markle type in the Hailes Member and mainly Dunsapie and Craiglockhart types in the East Linton Member. Varieties transitional between the macroporphyritic types have been recorded; microporphyritic Jedburgh types and transitions between the microporphyritic and macroporphyritic types are rare. In some flows, where phenocrysts are less abundant and/or irregularly distributed, individual specimens may not possess all the attributes of a specific type.

Mugearites occur in both the Hailes and East Linton members but 'leucite'-hornblende-kulaite occurs only at the base of the lava sequence.

Trachytes form the upper part of the volcanic sequence, the Bangley Member. Whereas the trachytes attain their maximum development and petrographical variation in the Haddington district, in the present district they are represented only by two small areas west and south-west of Whittingehame.

DETAILS

Basalts and hawaiites

Olivine-basalts of Craiglockhart type (ankaramites)

In general augite phenocrysts are more common than those of olivine. Both are commonly anhedral and corrosion is locally marked (S 10835); locally they are euhedral (S 11327). The larger olivines are normally represented by pseudomorphs; fresh olivine has been noted in only one rock (S 10835). The augite phenocrysts though locally chloritised (S 11303) are generally fresh and pale purple or brown in colour. Locally (S 11327) some of the augite phenocrysts are sieved with inclusions of iron ore and microlites of feldspar. Small well-shaped crystals of olivine or, more commonly, altered olivine occur in the matrix. The pale pyroxene of the matrix is finely granular or, in coarser specimens, occurs as small grains and prisms. In a fine-grained sliced rock (S 11327) from the Whittingehame area, roughly circular or ovoid patches of the matrix lack the normal feldspar microlites and there is an isotropic, analcimic base. Similar blotchy distribution of analcime has been noted in rocks of this type in the Haddington district. Specimens (S 10593, 10598, 10836) containing a few large stout tablets or small phenocrysts of plagioclase and with a matrix of more feldspathic appearance occur which may be regarded as transitional to Dunsapie type.

Olivine-basalts of Dunsapie type

The distribution and proportion of phenocrysts is variable. In some rocks (S 11309) the olivines or their pseudomorphs are present only as large microphenocrysts. Augite phenocrysts are absent from one sliced rock (S 631) from a flow at Kippielaw but are present in other specimens from the same area. Some rocks (S 11306, 11308) are extremely porphyritic.

The plagioclase phenocrysts are locally (S 53430) more plentiful than those of augite and olivine and such rocks trend towards Clark's (1956, p. 46) 'Feldspathic Dunsapie type'. The olivines of the phenocrysts and groundmass are normally altered and fresh olivine has been noted in only two sliced rocks (S 10839, 11339). The pale brown or purple augite of the phenocrysts and groundmass is generally fresh. The phenocrysts are locally sieved with inclusions and in one rock (S 52095) the margins of the phenocrysts have an ophitic relation with the groundmass feldspar. The augite of the matrix is granular or prismatic. The groundmass plagioclase is labradorite (An_{72} in S 55525), generally zoned and mantled by oligoclase and potash feldspar. Iron ore occurs as small grains and plates. The matrix may be coarse (S 11309, 54544) or fine-grained. A little intersertal analcime has been noted in a few sliced rocks (S 10834, 11309, 52093). A little brown hornblende and biotite locally occurs (S 52090).

Olivine-basalts and hawaiites of Markle type

Olivine-basalts of Markle type occur in the Hailes Member and the type locality (Markle Quarry) occurs in the district. The plagioclase phenocrysts are commonly fresh and anorthite values, determined by Carlsbad-Albite extinction values, range from An_{64} (S 10594, 11940) to An_{76} (S 629). The range of determined values probably indicates chance intercepts through zoned crystals rather than necessarily differing compositions in individual specimens. Oscillatory zoning is well developed in a specimen (S 11335) from Markle Quarry. Olivines are represented only by pseudomorphs. Small prisms of pale augite occur in the matrix. Potash-feldspar locally occurs interstitially (S 11347, 53415) and oligoclase mantles the more calcic plagioclase. Apatite is a common accessory as small needles and in a few samples (S 629, 53415) occurs also as quite large crystals. Specimens in which the feldspar phenocrysts are smaller (c.2 mm) and scattered may be regarded as intermediate between Jedburgh and Markle types (S 10596, 11342–43). Some varieties (S 53406) contain rare augite phenocrysts indicating a tendency towards Dunsapie type. One such specimen (S 10636) contains rare crystals of brown hornblende and some plates of foxy-red biotite of late crystallisation.

Microporphyritic olivine-basalts and hawaiites

Basalts of Jedburgh type are rare and have not been separated from the macroporphyritic basalts on the geological map and some have been mapped with the mugearites. One sliced rock (S 52106) from the Whitekirk area contains numerous large plagioclase microphenocrysts (up to c.1.5 mm long) indicating a tendency towards Markle type. It differs from the Markle type basalts and hawaiites of the district in having markedly ophitic (trachy-ophitic) plates of pale augite. Another specimen (S 53412) mapped with mugearite in the Lawhead area, contains lathy microphenocrysts of zoned labradorite (An_{66} at core) with pale purple augite as ophitic plates and as very elongate (up to c.1.8 mm long) subophitic plates. The calcic plagioclase of the matrix is accompanied by oligoclase and orthoclase. Another specimen (S 10597) from the same area contains laths of labradorite-andesine (An_{50}) zoned out to about oligoclase and with a fair proportion of alkali feldspar. The rock may be regarded as transitional to the mugearites and mineralogically as a hawaiite. Intersertal analcime occurs in a microporphyritic rock (S 11855) from Kippielaw. This rock is similar to a microporphyritic Jedburgh basalt but has some large olivine pseudomorphs and would, according to MacGregor (1928, p. 350) be regarded as intermediate between Dalmeny and Markle types.

The lavas (S 10697–10701) of the Oldhamstocks area, which (p. 36) are thought to be older than the Garleton Hills Volcanic Rocks, are extremely altered (argillised, chloritised and carbonated). Some specimens (S 10697, 10700) are of dolerite grainsize. The rocks are either non-porphyritic or may have pseudomorphs after olivine as small microphenocrysts.

Mugearites and trachybasalts

Mugearites

The rocks grouped under the heading mugearite probably cover a range of compositions from hawaiite through mugearite to benmoreite. The history of the usage of the terms in East Lothian has been discussed by Elliot (*in* McAdam and Tulloch, 1985). Some of the rocks (S 11344–45, 53416) contain scattered plagioclase phenocrysts. Commonly the groundmass feldspar is oligoclase-andesine and potash-feldspar may occur interstitially (S 10587, 10634); in some samples (S 10837, 11300) the feldspar is dominantly oligoclase. Flow-orientation of the feldspar is normally well developed. Pseudomorphs after olivine occur as small grains and locally as rare microphenocrysts (S 10634). Apatite is a common accessory mineral as delicate clear needles but locally as orange needles (S 11307) and small apatites striated with dark inclusions (S 53424) are locally common. Locally microphenocrysts of dusky apatite occur (S 10838). Grains of iron ore are common and may form the main dark mineral. Small grains of pyroxene occur locally. A rock (S 11345) grouped on the map with the mugearites in the Kippielaw area contains flow-aligned laths of oligoclase-andesine, interstitial potash-feldspar with slender prisms of pale pyroxene, small pseudomorphs after olivine and grains of iron ore. The rock

may be regarded as a hawaiite linked to trachybasalt. It contains rare flakes of foxy-red biotite with small plates of the late-stage brown hornblende (+ 2 V large) like that recorded from the Tertiary mugearites of Skye (Flett, 1908) and also in the Carboniferous mugearites, trachybasalts and more felsic basaltic rocks (Bailey *in* Clough and others, 1910, p. 124; MacGregor, 1928, p. 346; MacGregor *in* Richey and others, 1930, p. 107; Elliot *in* Francis and others, 1970, pp. 162, 164).

'Leucite'-kulaite ('leucite'-hornblende-trachybasalt)

The kulaite in the Whittingehame area contains numerous microphenocrysts of resorbed hornblende now represented by an aggregate of iron ore. Locally (S 50341) small relics of brown hornblende occur. Faintly purplish augite occurs as microphenocrysts and as small prisms. Clear analcime is common and very rarely small areas enclosing circular zones of inclusions are similar to those considered by Bennett (1945) to represent pseudomorphs after leucite. Some, large dusky apatites occur. In an altered specimen (S 50343) from the same area, though the texture is preserved, the matrix feldspar and analcime are argillised.

Trachytes

The trachytes of the Bangley Member show their maximum variation in the Haddington Sheet (McAdam and Tulloch, 1985) but in the Dunbar district are represented by only one sliced rock (S 50365). It is a fine-grained trachyte containing small orthophyric microphenocrysts of locally kaolinised orthoclase in a matrix of felted microlites of potash-feldspar, chloritic material possibly in part after ferromagnesian grains and small grains of iron ore. Good flow-structure and autobrecciation are developed.   RWE

CHAPTER 8

Intrusions

INTRODUCTION

Intrusions of various ages cut strata of all the solid formations in the district. The oldest are of late-Caledonian age,
ranging from felsite to granitic rocks, and cut Ordovician
and Silurian strata. Some evidence exists for the presence of
a major granitic batholith almost entirely concealed under
these rocks. Magmatic activity occurred several times during
the Carboniferous, intruding mainly Devono-Carboniferous
and Carboniferous strata and also the Lower Palaeozoic
sediments to a lesser extent.

INTRUSIONS OF LATE-CALEDONIAN AGE

During late Silurian and possibly early Lower Devonian
times minor resurgences of the Caledonian Orogeny occurred and they were accompanied by the intrusion into
Lower Palaeozoic strata of dykes of quartz-porphyry, acid
porphyrite and porphyrite, and less commonly plagiophyre,
microgranodiorite and lamprophyre. The Priestlaw and
Cockburn Law plutons were also emplaced during this
period.

The Silurian rocks of the district are cut by numerous
dykes of late-Caledonian age. For the main part the dykes
have trends approximately parallel to the country rock and
therefore to the west of the Great Conglomerate the trend of
the dykes is mainly NE–SW whereas in the eastern part of
the district the strike of the country rocks and trend of dykes
is nearer N–S.

It is possible that the rather altered quartz-porphyry dykes
may include intrusions of Carboniferous age since the Carboniferous felsite (or quartz-porphyry) intrusion of Dirrington Law occurs in the area. A few dykes of badly altered
rocks are also present, the age of which is uncertain.

The minor intrusions are, in the main, extremely altered;
the mafic minerals are normally represented only by
pseudomorphs and the plagioclase is commonly highly
sericitised and albitised. Replacement of feldspars by
kaolinite and carbonate also occurs. As in the similar intrusions of the Haddington district (McAdam and Tulloch,
1985), many of the dykes within the aureoles of the
granodiorite masses are fresher. It may be that here also, as
suggested by MacGregor (in Richey and others, 1930,
pp. 35, 51) in connection with the Distinkhorn Complex,
there is a pre- or early-metamorphic alteration. MacGregor
attributed this effect to circulating heated water and vapours
prior to the intrusion of the complex. The difference in
degree of alteration may, however, be caused in part by
hydrothermal solutions younger than the granite and may
reflect the greater ease of circulation of fluids in the country
rocks as opposed to in the granites or their aureoles.

DETAILS

Priestlaw intrusion

The Priestlaw intrusion is an irregularly-shaped pluton lying to the
north-west of Priestlaw Hill [653 624] and intruded into Silurian
rocks. It occupies the lower ground of the valleys of the Whiteadder
Water and its tributary the Faseny Water. Part of the area of outcrop is covered by the Whiteadder Reservoir. The area of outcrop is
about 5 km in a WNW to ESE direction and about 2 km from NNE
to SSW at the maximum width of outcrop.

Exposures of the solid rock are seen in the banks of the Whiteadder and Faseny waters. The intrusive relationship with the
Silurian strata is seen in the Faseny Water about 1.6 km up from
the junction with the Whiteadder although the actual contact is not
seen. Elsewhere, where the rock is exposed it has decomposed into a
yellowish or greenish sand with nodular lumps of solid rock in a
loose matrix.

The rock has been identified as a biotite- or hornblende-biotite-
granodiorite and the grain size increases from the periphery to the
centre of the intrusion. The pluton has been described by Geikie (in
Howell and others, 1866, p. 15), Peach and Horne (in Clough and
others, 1910, pp. 22–23) and Walker (1925; 1928).   RWE

Cockburn Law intrusion

Part of the outcrop of the Cockburn Law intrusion occurs in the
south-east corner of the district, around Cockburn Law [765 597].
The larger proportion of the outcrop lies to the east, in the
Eyemouth district (Greig, in press).

The intrusion is a small, irregularly-shaped pluton which
measures roughly 2 km from north-west to south-east and 1 km
from south-west to north-east. The south-west margin is linear and
many be determined by a fault. The rocks range in composition
from diorite at the margins to granodiorite at the centre. They are
intruded into deformed greywackes and shales of Llandovery age
and are overlain unconformably to the east of the district by Upper
Old Red Sandstone strata which contain pebbles of granodiorite
and hornfelsed greywacke.

The rocks are poorly exposed and rather altered. Exposures
around Cockburn Law, near the north-west margin of the intrusion, consist of dark grey quartz-diorite and the rock becomes paler
in colour towards the south-east. Central parts of the intrusion
which are composed of granodiorite, are exposed in the banks of the
River Whiteadder just to the east of the district margin. The contact
with the country rocks is not exposed. Thermal metamorphism of
the greywackes and shales occurs in a zone which is about 300 m
wide on the south-west side of the intrusion, but the zone is up to
1 km wide to the north and north-east. The intrusion has been
described in papers by Walker (1925; 1928) and Midgley (1946)
and an isotopic age of 408 ± 5 Ma (Rb-Sr) is given by Brown and
others in Harris (1985).

The concept of a granitic batholith under the Southern Uplands
was mentioned by Midgley (1946) and the Cockburn Law intrusion
was postulated as part of the Tweeddale Granite which was considered, on geophysical evidence, to underlie much of the Southern
Uplands (Lagios and Hipkin, 1979).

Dykes

Petrography

Acid porphyrite

Acid porphyrite is the most common rock-type of the late-Caledonian dykes which were intruded into deformed Lower Palaeozoic greywackes and shales, but do not penetrate the Lower Devonian conglomerates. The dykes have a general north-east to south-west trend and are often concordant in steeply dipping strata. They range in thickness from 1 m up to 24 m but they cannot be traced laterally from one stream section to the next. The rocks are buff, pink or red-brown in colour, fine- or medium-grained and porphyritic.

Good examples of acid porphyrite dykes may be seen in the Monynut Water and the Whiteadder Water east of the Lower Devonian conglomerate outcrop.

Quartz-porphyry

A quartz-porphyry dyke about 3 m thick with a north–south orientation occurs near Nether Monynut [7283 6425]. It is a fine-grained, pink rock with phenocrysts of quartz in a quartz-feldspathic matrix. It occurs associated with an acid porphyrite dyke.

Porphyrite

Two porphyrite dykes are known to occur in the eastern part of the area. They have the same orientation and mode of occurrence as the acid porphyrites. One dyke 1.2 m wide, occurs in an old quarry [7602 5834] south-west of Cockburn Farm, and the other, which is about 24 m wide, crops out in the Whiteadder Water [7645 6155] near Abbey St Bathans. The rock is fine- to medium-grained and consists of phenocrysts of intermediate plagioclase, hornblende, pyroxene or biotite in a matrix of these minerals with some iron ore, and it is usually rather altered. Two porphyrite dykes on 1:50 000 Sheet 34 (Eyemouth) were found to be thermally altered in the vicinity of the Cockburn Law Intrusion, but this is not true of all late-Caledonian dykes.

Lamprophyre

There are two augite-minette dykes in the east of the district, one exposed in the Whiteadder Water [7458 6136] near Barnside and the other in the Monynut Water [7480 6296] near Godscroft. Another unspecified lamprophyre crops out in the Monynut Water [7493 6289] near Strafortane Mill.

The lamprophyres occur as irregular discordant dykes or small masses from 0.2 m wide up to 4.5 m wide with a north-east to south-west trend.

They are intruded into deformed Lower Palaeozoic rocks and although none are known to cut Lower Devonian strata in this area such a relationship is known from the Berwickshire coast. The rocks are dark reddish-brown or grey in colour, fine-grained and commonly more or less altered.

Microgranodiorite

Three minor intrusions of microgranodiorite are known in the district. One is a dyke about 3 m wide with a north-east to south-west orientation and the other two are small bosses up to 22 m across. The rock is fine- to medium-grained and is pink or reddish-brown in colour.   IBC

Quartz-feldspar-porphyry

A number of dykes of quartz-feldspar-porphyry occur. They are characterised by generally rounded, corroded phenocrysts of quartz (S 49365) accompanied by phenocrysts of sodic plagioclase sericitised to varying degree (S 49372–73, 49388–89, 52866) and scattered phenocrysts composed of an aggregate of white mica and hematite probably pseudomorphing biotite. In a number of specimens (S 49366, 49387–89, 50335) phenocrysts of generally turbid potash-feldspar, some microperthitic, occur. In one specimen (S 49387) sericitised plagioclase phenocrysts are mantled by a thin layer of clearer potash-feldspar. The matrix of the porphyries is commonly a granular mosaic of quartz and alkali feldspar with a fair proportion of small flakes of sericite, the latter being locally plentiful. Locally micropoikilitic plates of quartz occur (S 49365). The relative proportions of the phenocrysts are variable. In some rocks (S 49361, 49371, 49375) the feldspar phenocrysts are only sodic plagioclase but in others potash-feldspar also occurs. In view of the occurrence of the Lower Carboniferous felsite (and quartz-feldspar-porphyry) intrusions at Dirrington Law it is possible that some of the dykes cutting the Silurian rocks are of Carboniferous age. However they differ from the felsites of Dirrington Law in that they are all characterised by phenocrysts of sericitised sodic plagioclase which is relatively uncommon in the Carboniferous intrusion and the proportion of phenocrysts of potash feldspar is generally smaller.

Acid porphyrite

The acid porphyrites contain numerous phenocrysts of albite-oligoclase, biotite and, in some instances, hornblende. The ferromagnesian minerals are represented only by pseudomorphs in white mica, chlorite, carbonate and hematite. Partly resorbed phenocrysts of quartz occur in a few specimens. The matrix is composed of small plates and crystals of alkali feldspar, local laths of sodic plagioclase, and much quartz, normally with many flakes of sericite or white mica. The texture of the matrix is variable. In some rocks (S 50328–29, 50340, 52859) the matrix is a granular mosaic, coarse or fine, of quartz and alkali feldspar; in others the quartz occurs as large poikilitic plates (S 49393, 52842, 52846). Quite extensive microgranophyric intergrowths occur in some specimens (S 49394, 50340) and locally spherulites of quartz and alkali feldspar are present (S 49393, 49395, 52842). The plagioclase is commonly sericitised, sometimes intensely so (S 48195, 49380, 52842), but is generally albite-oligoclase and the phenocrysts in particular may represent albitised plagioclase. Some of the rocks, particularly those with a granular matrix, might be described as feldspar-porphyries.

Microgranodiorite

Two small intrusions of microgranodiorite or fine-grained granodiorite (S 46650, 48181) have been recorded in the Monynut Water area. They contain phenocrysts and plates of sericitised, zoned plagioclase, flakes of locally chloritised biotite and anhedral quartz. Potash-feldspar mantles the plagioclase and some patches of delicate micropegmatite occur. In one specimen (S 46650) there are pseudomorphs in carbonate and chlorite, possibly after hornblende.

Granodiorite

In addition to the large Priestlaw and Cockburn Law intrusions a 27-m wide intrusion of coarse granodiorite occurs on the south side of Monynut Water. The rock (S 46645) is composed of coarse crystals of generally intensely sericitised, highly-zoned plagioclase

(An$_{54}$ at core), locally showing oscillatory zoning, plates of locally chloritised biotite, pseudomorphs of chlorite and carbonate probably after hornblende, potash-feldspar mantling the plagioclase, and anhedral quartz.

Lamprophyre

The few lamprophyre intrusions are mainly augite-minettes typically containing rare pseudomorphs after olivine, numerous carbonate pseudomorphs after euhedral augite and many dark-rimmed tablets and flakes of brown biotite. The augite is locally fresh (S 46649, 52863). Olivine pseudomorphs are locally plentiful (S 52863) or may be apparently absent (S 49361). The feldspar is dominantly hematite-dusted, slightly sericitised, potash feldspar which occurs generally as small laths and plates but may be coarse and anhedral (S 46642, 46649) or rarely forms a fine-grained, feathery aggregate (S 49361, 52863). Iron ore occurs as plates and rods. Quartz occurs intersertally and also (S 604, 46642, 46649), as xenocrysts. Feldspathic ocelli may be present. A dyke of biotite-bearing spessartite (S 46647) has been recorded in which many flakes of brown biotite accompany numerous pseudomorphs after euhedral hornblende. The matrix is composed of somewhat sericitised, zoned plagioclase with a little intersertal quartz.

Porphyrite

The porphyrites range from relatively fresh rocks, in which at least some original minerals are preserved, to extremely altered rocks (S 50335–37, 50539) in which the ferromagnesian minerals are replaced by chlorite and the feldspar commonly largely or completely sericitised. The rocks are characterised by phenocrysts of plagioclase and phenocrysts and microphenocrysts of pale brown or green hornblende (S 52804, 52809, 52848) and locally of biotite (S 52865). The plagioclase is commonly somewhat sericitised, acid plagioclase but this is apparently albitised and relicts of more calcic plagioclase are seen in some phenocrysts (S 50385, 52809). Locally, later kaolinisation has preferentially attacked such relicts of fresh plagioclase (S 56385). Fresh hornblende is locally seen and also phenocrysts of foxy-red biotite. These minerals are normally represented by pseudomorphs. Rarely (S 52809, 52812) chlorite pseudomorphs, possibly after orthopyroxene, occur.

Thermally altered dyke rocks in the aureoles of the Priestlaw (S 7811, 49386, 50326, 50354, 50356) and Cockburn Law (S 48173, 48176, 48199) masses contain phenocrysts of plagioclase locally (S 49386) as calcic labradorite (An$_{66}$ at core) and with discontinuous zoning. The plagioclase may be sieved by inclusions and locally turbid but also commonly shows pronounced thermal clouding (S 49386). New thermal biotite occurs speckling the granular, generally recrystallised groundmass and also as decussate aggregates, some replacing hornblende or replacing chloritised hornblende. Locally fibrous aggregates of pale green amphibole occur with biotite and some may be after pyroxene. Other dyke rocks outwith the immediate aureole of the Priestlaw pluton are somewhat altered and, in several, minerals of the epidote group partly replace the feldspar (S 50355, 52803, 52809, 52812) or pseudomorph biotite (S 50355, 52796, 52812, 52848); epidotisation also occurs around the Cockburn Law mass.

There are two generations of dykes, one pre-dating the granitic masses and thermally altered, and the other set post-dating and cutting the granite (S 58324) or cutting the thermal aureole (S 7812) and this is in accord with dyke–granite relationships in the Haddington district.

Plagiophyre

A few very altered, effectively non-porphyritic dyke-rocks have been grouped as plagiophyres as defined by MacGregor (1939, p. 101). They are composed of laths of turbid albitised plagioclase, in some (S 49369) as a plexus, in others (S 49382, 49396) flow-orientated, and quartz may occur as intersertal or spongy areas. The feldspar is locally highly sericitised (S 49383). Small pseudomorphs after poorly shaped ferromagnesian mineral occur.

Priestlaw Granodiorite

As noted by Walker (1925; 1928) the Priestlaw intrusion has a marginal phase of slightly porphyritic quartz-augite-biotite-diorite or microdiorite giving place, towards the centre of the mass, to a non-porphyritic granodiorite and hence to a centre of pink porphyritic hornblend-biotite-granodiorite. An area of olivine-norite has been mapped by M. F. Howells at the north-west corner of the Priestlaw mass.

The granodiorite contains phenocrysts of calcic plagioclase, some with marked oscillatory zoning (S 7809, 56322) with green hornblende and brown biotite in a matrix generally of finely granular quartz and feldspar which in one specimen (S 7808) is relatively coarse. Most of the other rocks of the complex may be regarded as fine-grained quartz-diorite or as slightly porphyritic quartz-microdiorite. One specimen (S 56321) is a fine-grained quartz-hornblende-biotite-diorite with a little potash feldspar. In addition to biotite the rocks are mainly characterised by the presence of pale clinopyroxene (S 7810, 50323). In one rock (S 50330) the pale clinopyroxene is commonly mantled by green hornblende. A specimen (S 50367) from Penshiel contains biotite, green hornblende and a little pale clinopyroxene, the latter locally mantled or replaced by hornblende. A coarse dioritic rock (S 50331) from Kell Burn contains large crystals of deep red-brown hornblende in addition to biotite, and aggregates of biotite occur suggestive of thermal metamorphism.

The olivine-norite (S 50389) contains sparse anhedral grains of olivine, largely replaced by bowlingite, coarse subhedral prisms and plates of orthopyroxene and clinopyroxene, very coarse plates of zoned calcic plagioclase, anhedral plates of iron ore and sparse intersertal quartz. A little turbid alkali-feldspar mantles the plagioclase and a few flakes of biotite occur. RWE

INTRUSIONS OF CARBONIFEROUS AGE

At various times during the Carboniferous, magmas penetrated the Devono-Carboniferous and Carboniferous rocks of the district and consolidated into various types of intrusive bodies. Their form and relationships with the country rocks are best known where they are exposed on the coastal section. Other intrusions may be concealed under the drift-cover inland. A few minor intrusions of Carboniferous age cut the Lower Palaeozoic rocks. The Dirrington Law felsite and a suite of basic dykes cut the Lower Devonian conglomerate.

Many of the intrusions are sill-like bodies. Volcanic plugs are present, such as the Bass Rock (Plate 10) and St Baldred's, and a laccolith occurs at Traprain Law (Plate 13). The dykes take many forms including swarms and small irregular networks associated with vents.

Classification

The intrusion can be divided into groups depending on their lithology and age:

Plate 10 The Bass Rock, phonolitic trachyte plug (MNS 1389)

Late Westphalian to Stephanian	Quartz-dolerite and tholeiite dykes
?Namurian	Olivine-dolerite/teschenite suite
?Dinantian to Stephanian	Monchiquite/basanite suite
Dinantian (Garleton Hills Volcanic Rocks)	Trachytic intrusions
Dinantian (Garleton Hills Volcanic Rocks)	Basaltic intrusions

Age of intrusions

The evidence concerning the various ages of the intrusions has been given in the memoir on the adjacent Haddington district (McAdam and Tulloch, 1985). A few intrusions are related in form and lithology to the Dinantian Garleton Hills Volcanic Rocks. The basalt sill at Primrose Bank and the St Baldred's Plug are in this category, although the latter has given a Westphalian age-date (McAdam and Tulloch, 1985, table 3). The trachyte sill at Garvald is lithologically similar to the lavas, but the phonolitic trachyte of the Bass Rock Plug and the Traprain Law Laccolith are not found as lavas.

They are thought to have been produced at a late stage after the eruptive phase of advanced differentiation by the alkaline magma (Tomkeieff, 1937; MacDonald, 1975). The Dirrington Law felsite has an age-date also suggesting late-Dinantian. Other late stage differentiates, not found as lavas, are minor intrusions of the monchiquite-basanite suite, associated with vents along the coast such as The Car and the Dunbar vents. Teschenite sills at Ravensheugh and Frances Craig are probably related to the Namurian sills found in other parts of central Scotland, as indicated by an age-date for the former. The W–E quartz-dolerite dyke at Dunbar and those in the south-east of the district are part of the late-Westphalian to Stephanian quartz-dolerite suite of the Midland Valley.

Numerous occurrences of xenoliths of ultrabasic igneous rocks and, to a lesser extent, of metamorphic rocks have been recorded in the Haddington district (McAdam and Tulloch, 1985). In the Dunbar district Upton and others (1983, pp. 112, 113) have recorded xenoliths of quartzo-feldspathic gneiss and also glimmerites at Beggar's Cap and Dunbar. They also recorded megacrysts of mica in the Dunbar vent probably produced by disaggregation of glimmerites.

Plate 11   Hexagonal cooling joints in basanite plug, The Battery,
Dunbar Harbour (D 3660)

DETAILS (Letters in brackets refer to Figure 10)

Basaltic intrusions related to the Garleton Hills Volcanic Rocks

St Baldred's Plug (SB)

Almost completely filling the Whitberry Vent (Day, 1930a,
pp. 230–232) is an oval (300 × 200 m) plug of fresh Craiglockhart
type basalt forming Whitberry Point and St Baldred's Cradle. It is
intruded into bedded sandstones and red marls of Calciferous Sandstone Measures age which form the foreshore, and the junction with
the sediments is at or near H.W.M. for over half the circumference
of the plug. The junction of basalt with red vent-tuff can also be
seen in the south-east. The dip of the country rock increases from
10° or less to about 45° inwards at the plug margin possibly due to
collapse after intrusion. The rock is black, fresh and well jointed,
containing large black augite phenocrysts in a fine-grained, basaltic
matrix.

Primrose Bank Sill (PB)

This sill (Day, 1930a, pp. 218–220) of Craiglockhart type basalt
crops out in the cliff behind the post-Glacial raised beach. It is intruded concordantly into low-dipping red-stained, bedded, basaltic
tuffs of the North Berwick Member, which are baked. At the

eastern end of the outcrop a block of country rock, 3 m long, is
caught up near the base of the sill. The rock is similar to that of St
Baldred's Plug.   ADM

Petrography

Feldspar-phyric olivine-basalts allied to Markle type form the intrusions of Kirklandhill [618 779] and at The Battery and Meikle
Spiker, Dunbar. The Kirklandhill rock (S 53405), which is
analcime-bearing, has many phenocrysts of olivine with some
microphenocrysts of augite and is transitional to Dunsapie type.
The plagioclase of the Meikle Spiker rock (S 11312–13, 61158) is
fresh but that of The Battery intrusion (S 998, 47040) is kaolinised
and carbonated. Analcime-bearing basalts allied to Craiglockhart
type (ankaramite) form the intrusions of Primrose Bank (S 10833)
and St Baldred's Cradle (S 10792). Upton (*in* Sutherland, 1982,
p. 515) recorded phenocrysts of opaque spinel in the St Baldred's
Cradle rock.   RWE

Trachytic and other felsic intrusions related to the Garleton Hills Volcanic Rocks

Garvald Sill (Ga)

A sill of trachyte caps the high ground west and south of Garvald
from Law Knowes towards Overfield, and is seen only as isolated

knolls. The sill is presumably underlain by Devono-Carboniferous red sandstones but the contact is everywhere obscured. The dip of the sill is probably low, and it is uncertain whether the top cover is anywhere preserved.

Bass Rock Plug (Ba)

Undoubtedly the most famous topographical feature of the Dunbar district the offshore islet of the Bass Rock (Clough and others, 1910, pp. 97, 129–130) has been long known to geologists as a classical volcanic plug and is a widely visible landmark. Its vertical sides reaching from over 45 m above to at least 18 m below OD reflect the edge of the resistant plug. The grey rock is hard and massively jointed and very fresh under a superficial weathering. It is composed of non-porphyritic, orthophyric, fairly coarse-grained, phonolitic trachyte.

Traprain Law Laccolith (Tr)

This intrusion of phonolite (Plate 13) was first described by Bailey (*in* Clough and others, 1910, pp. 98, 128–129). Part of it crops out on the western margin of the district and the western part of the laccolith is in the adjoining Haddington district. A full description of the intrusion, including the details of its structure and petrography have been given in McAdam and Tulloch (1985, pp. 62–65).   ADM

Dirrington Great Law intrusion

Dirrington Great Law is a prominent conical hill at the southern margin of the district [698 549] and is composed of felsite. It is intruded into Lower Devonian conglomerate and is probably overlain by strata of Upper Old Red Sandstone facies although the relationships are not clear in the field. The form of the intrusion is thought to be laccolithic. Exposure is poor but samples from a number of small quarries and loose angular debris in the area of outcrop show little variation in rock-type. An age of 325 ± 10 Ma (K-Ar) has been obtained for the intrusion which suggests approximate contemporaneity with the Garleton Hills Volcanic Rocks.   IBC

Petrography

Trachytic rocks   The Garvald trachyte intrusion (S 11240, 11243) is a non-porphyritic trachyte in which the laths of alkali feldspar show good trachytic flow-alignment.

Campbell and Stenhouse (1934) have given a detailed account of the petrography of the phonolitic trachyte of the Bass Rock and first recorded in it the presence of nepheline and fayalite and the probable occurrence of sodalite. The rock is composed dominantly of alkali feldspar in lath-shaped or orthophyric crystals of varying length with lesser amounts of anhedral aegirine-augite, fayalite, nepheline and magnetite with analcime and sodalite in intersertal or drusy patches. The yellow fayalite (S 10795) is largely altered to chlorite or chlorophaeite. Nepheline is generally pseudomorphed by turbid analcime (S 56628). Flow structure is locally well developed. Campbell and Stenhouse recorded the presence of cognate xenoliths of soda syenite.

The petrography of the Traprain Law intrusion has been described by Elliot (*in* McAdam and Tulloch, 1985).

Felsite   The felsite (or quartz-porphyry) of Dirrington Great Law is similar to those of the Dirrington Little Law and Blacksmill Hill intrusions just south of the district. They were described by Irving (1930) who indicated similarities to the Carboniferous intrusions of the Eildon Hills.

The felsite contains phenocrysts of quartz alone (S 1659) or, more generally, accompanied by potash-feldspar. Phenocrysts may be rare (S 52862) or numerous (S 33905, 49157). In many specimens the potash-feldspar phenocrysts are replaced by coarsely crystalline kaolinite. A few large flakes of biotite occur (S 1654, 13579, 49157). Rarely (S 49158) the orthoclase mantles a core of sodic plagioclase. Irving (1930, p. 534) noted this and recorded albite phenocrysts near the summit of Blacksmill Hill. The matrix is generally a fine-grained, quartzo-feldspathic aggregate of granular texture (ED 6635, S 52862) but micrographic texture (S 13579) is locally developed and, in some samples (S 1654, 49158) small rectangular laths of orthoclase occur. Biotite and soda-amphibole (pleochroic in orange and green) occur as scraps in the matrix of fresher specimens from the Dirrington laws (ED 6635, S 1656–57). In many specimens the matrix is silicified and altered to a quartz-mosaic (S 1660–61, 33903). Fluorite occurs in rocks from Dirrington Great Law (ED 6635, S 1654, 33905–6) and Dirrington Little Law (S 1655) in veinlets associated with quartz and kaolinite and also partly replacing feldspar phenocrysts.   RWE

Basanite-monchiquite suite

The Car dykes (TC)

A lenticular, black, leucite-basanite dyke, up to 2 m across, cuts The Car Vent near the Car Beacon. An irregular basanite intrusion and a thin dyke lie 150 m to the south-east, and basanite forms St Baldred's Boat (Day, 1930a, pp. 213–233).

Taking Head Sill (TH)

Within the Gin Head Vent (p.37) there is a large irregular monchiquite sill (Day, 1928, pp. 45–46) at least 10 m thick, well exposed on the foreshore, in stacks such as Tapped Rock, and in the cliffs between Canty Bay and Oxroad Bay. The rock is very hard, fresh, black, sparsely porphyritic, closely jointed and deeply weathered in places along the joints. On the shore at H.W.M. 50 m due north of Taking Head are irregular dykes representing a late stage in the intrusion.

Seacliff Plug (SE)

A plug of Hillhouse basalt occurs in a small vent which cuts red-stained tuffs (Day, 1930a, pp. 215–216). It is exposed in the Seacliff Quarry and adjacent cuttings where up to 5 m of blue-grey, fine-grained, non-porphyritic basalt are exposed. It has strong horizontal jointing which is flexured into gentle folds, possibly as a result of vent-processes. One section shows 0.6 to 1.2 m of red-purple, unbedded tuff possibly vent-tuff, occurring between the basalt and the bedded tuff of the country rock. Celestine veins are an unusual feature of the plug.

Stenton Sill (ST)

Several isolated sections of basanite near Stenton represent a dissected sill intruded into basaltic tuff and limestones of the North Berwick Member along the Dunbar–Gifford Fault. Exposures occur in valleys in two places in the Whittinghame Water–Luggate Burn area where up to 5.5 m of basanite are exposed, and in the Stenton Burn at Ginglet where 8.5 m of this rock occurs in a quarry. The rock is blue-grey with brown phenocrysts in a fine matrix.   ADM

Petrography

In the Midland Valley the term monchiquite has been applied incorrectly to rocks lacking hornblende and the term has been applied to feldspathoid-rich, pyroxene-rich rocks probably more correctly described as mafic olivine-analcimite or olivine-nephelinite. The basanites and 'monchiquites' are generally characterised by phenocrysts or microphenocrysts of olivine and augite and the matrix is commonly pyroxene-rich.

Basanite of the Kidlaw type of the Haddington district forms the Knockenhair Plug of Dunbar (S 11317, 61161). Biotite is common and analcime forms large areas in the matrix. Analcime also occurs as large ovoid areas, relatively free from feldspar, in a specimen (S 57003) of nepheline-basanite from the Taking Head intrusion. Similar analcime areas free of feldspar occur in the analcimic basalt or basanite (allied to Hillhouse type) of the Seacliff Plug (S 21430, 28300, 29621).

'Monchiquites' or olivine-analcimites relatively free from feldspar include the dyke (S 11354) west of Presmennan Loch and monchiquitic' rocks relatively free of feldspar (S 1003, 11353) occur as varieties of some basanite intrusions such as those at Taking Head and Stenton. The texture of feldspar may vary in the basanites. In the Stenton intrusion the plagioclase may form laths (S 11237) or large poikilitic plates (S 50381).

Pseudomorphs after nepheline form poikilitic plates in specimens from the Dunbar Castle basanite intrusion (S 11314) and in the Garvald Mains 'monchiquite' Plug (S 11350).

The unusual leucite-basanite intrusions of The Car Vent were described by Balsillie (1936). Small analcimised leucites occur in a specimen (S 36778) from the dyke near Car Beacon and in specimens of the eastern intrusion they are plentiful (S 32880) or apparently absent (S 54098). Olivine is not plentiful and one specimen (S 32879) is more correctly described as a tephrite. Rounded microphenocrysts of brown hornblende (S 32879–80, 36778, 54098) occur in The Car Vent intrusions.

The Fernylees intrusion, near Oldhamstocks, is a coarse analcimic basalt allied to Hillhouse type (S 50311–16) but locally the presence of augite and olivine phenocrysts (S 10702, 50309) indicate a link with Craiglockhart type. Similarly the Seacliff Plug rock, an analcimic basalt or basanite allied to Hillhouse type, has microphenocrysts of olivine (pseudomorphs), augite and iron oxide (S 21430) but locally larger pyroxenes occur (S 28300) indicating a trend to Craiglockhart type.

An unusual vein noted by Day (1928, pp. 45–46) cuts the Taking Head Sill. It (S 54092, 57002) contains euhedral crystals of pale augite zoned to purple with smaller, often skeletal, crystals of deep purple augite, stout laths of plagioclase and rods of iron ore in a pervading glassy base and rare pseudomorphs after olivine.   RWE

Olivine-dolerite and teschenite suite

Ravensheugh Sill (RA)

Two upstanding masses of teschenite at Ravensheugh Wood are taken as dissected parts of the same sill (Day, 1930b, p. 261). There is no evidence of contact with the country rock, though intrusion-breccia and baked sandstone, possibly a xenolith, are present below the teschenite in the east at H.W.M. The teschenite is well exposed along the post-Glacial raised beach back-feature to the north and east of both masses, where it has also been quarried, but the boundaries to the south and west are obscure. The intrusion is at least 15 m thick, and is cut to the east by a small fault. The rock is dark grey and well jointed.

Frances Craig Sill (FC)

On the foreshore a small sill with a complicated intrusion-pattern is exposed. It dips gently to the west and is intruded into sandstone. The brown, very altered basic rock was described as a mugearite by Bailey (in Clough and others, 1910, p. 97) but Day (1930b, pp. 260–261) thought it more likely to be an altered version of the Ravensheugh teschenite.   ADM

Borthwick Sill

In the south-east corner of the district there are three outcrops of olivine-basalt which are poorly exposed but form low hills near Chapel Farm [765 574], Castle Mains [764 560] and Harelawcraigs Plantation [765 552]. The largest of the outcrops, at Harelawcraigs Plantation, includes Borthwick Quarry [770 540] which lies just beyond the southern margin of the district.

The outcrops are thought to be part of a large, sub-horizontal sill, probably of Carboniferous age, intruded into Devono-Carboniferous sediments. The rock is a black, fine- to medium-grained olivine-basalt or olivine-dolerite with analcime. The sill appears to thin towards the north from a thickness of at least 20 m at Borthwick Quarry.   IBC

Petrography

The only fresh olivine-analcime-dolerite noted is that of the Ravensheugh Sill. The rock contains a few pseudomorphs after olivine, pale subophitic augite, laths of zoned plagioclase and plates of iron ore. Flakes of biotite occur and analcime is present intersertally and in vesicles and partly replacing feldspar. The rock of the Frances Craig Sill (S 10794, 36786) is generally a rather altered olivine-dolerite but Day (1930b, pp. 260–261), probably on the basis of fresher material, considered the rock to be a teschenite.

The Borthwick Sill, exposed in Borthwick Quarry, is composed of coarse-grained, analcime-bearing olivine-basalt or dolerite (S 33908–9, 34689, 48233) containing numerous crystals of fresh olivine, laths of zoned calcic plagioclase and purple-brown augite as small, commonly aggregated, prisms and rarely as phenocrysts. Alkali feldspar occurs intersertally, and locally intersertal patches of analcime are found. The iron ore occurs as plates, and, characteristically, as stout rods.   RWE

Quartz-dolerite suite

Quartz-dolerite dykes penetrating Llandovery strata are seen in the Dye Water [632 580] near Byrecleuch, where two NE-trending dykes occur. Further quartz-dolerite dykes are intruded into Dinantian strata at Belhaven Bay [662 793] and into the Parade Vent [671 796] at Dunbar. A quartz-dolerite dyke, 30 m wide, trending north-east cuts Lower Limestone Group beds at Millstone Neuk [709 780] on the coast 3 km east of Dunbar, and has been exposed in the Dunbar Quarry.   AD

A dyke of quartz-dolerite aligned west to east, crops out near Cockburn Farm [767 586]. The dyke is 15 to 18 m wide, about 2 km long, and consists of dark grey or black quartz-dolerite intruded into Lower Palaeozoic greywackes and shales. It also cuts the Cockburn Law intrusion. The quartz-dolerite dykes are thought to be of late-Carboniferous age. Exposures of quartz-dolerite also occur in the Dye Water [708 586] and in the Whiteadder Water [699 599].   IBC

Petrography

The dykes at Belhaven are mainly altered tholeiites (in the petrographical sense) or tholeiitic dolerites with an intersertal glassy mesostasis. In the southern dyke (S 754, 11318) the pyroxene and plagioclase are altered. An unusual variety (S 11316) from the north dyke is coarse, has fresh plagioclase in a plentiful glassy mesostasis, and lacks clinopyroxene. The Millstone Neuk dyke (S 61155) is a carbonated tholeiitic rock with fresh plagioclase. In the southern part of the district a few dykes of the suite occur. Some are coarse (S 52851) or very coarse (S 52868) quartz-dolerites. Even where the plagioclase may be fresh (S 48227, 52851) only rarely (S 48220, 52868) is residual unaltered pyroxene preserved. In the tholeiites (S 49404) ocellar areas occur and in chilled varieties (S 52853) iron

Plate 12   Vertical dolerite dyke with close joints cutting Lower Devonian conglomerate, the baked conglomerate proving more resistant to erosion, Fairy Castle, junction of Back Water and Bladdering Cleugh (D 2746)

ore crystallites occur as a reticulate mesh in the decomposed glassy base.   RWE

Other intrusions

A swarm of dykes of olivine-basalt or olivine-dolerite cut the Lower Devonian Great Conglomerate, and are indeed confined to the conglomerate. These dykes, which may be of Middle Devonian or possibly of Carboniferous age, are generally considerably altered, and are mainly confined to the northern half of the outcrop. The dykes are generally less than 1 m thick, but range up to 5 m. Most of these intrusions are vertical, with a few dipping steeply, and all are characterised by sharp, clean junctions with the country rock (Plate 12). The greywacke clasts are cut sharply across, and the baking of the immediately adjacent conglomerate often forms clean-sided trenches resulting from the weathering of the less resistant dyke rock. The trend of these dykes varies in an arc from 270° to 325°, with a dominant trend of 300° to 305° (Figure 13). A few unclassified dolerite, basalt or tholeiite dykes also occur, which are also orientated within the azimuths described above.   AD

Petrography

Apart from a few dykes of the Stéphanian quartz-dolerite suite, the dykes, thought to be of Carboniferous age, are largely extremely altered olivine-basalts. Olivine and augite are represented only by pseudomorphs, and plagioclase, rarely fresh, is kaolinised and albitised. In one specimen (S 52792) the originally more calcic cores of zoned plagioclase have been preferentially kaolinised. Some rocks (S 52882) are completely altered to carbonate, kaolinite and red iron oxide. Rather feldspathic ocelli occur locally (S 50301, 50317, 52772, 52793); biotite may occur in the ocelli (S 46635) or as scattered flakes in the matrix (S 50290, 52772) and is locally (S 50292) accompanied by scraps of kaersutite. Analcime and zeolite occur in amygdales in one specimen (S 52777). Most of the basalts are fine-grained olivine-microphyric types (S 52773,

52779–81) rarely with lathy microphenocrysts of altered feldspar (S 52819). Other varieties of basalt include a coarse basalt (S 50317) with phenocrysts of kaolinised plagioclase and less common, large pseudomorphs after olivine and possibly augite. A relatively fresh rock (S 50300) has many lathy microphenocrysts of zoned labradorite (An_{62} at core) and microphenocrysts of pseudomorphed olivine and augite. Other specimens (S 50290, 52777) contain many pseudomorphs after olivine and augite as phenocrysts and microphenocrysts.

RWE

Plate 13   Traprain Law phonolite laccolith, from south-east (MNS 1449a)

CHAPTER 9

Structure of the post-Silurian rocks

INTRODUCTION

It is possible to relate the faulting and folding seen in the district to several phases of earth-movements.
1 Caledonian Orogeny. Pre-Lower Devonian (Lower Old Red Sandstone) earth-movements described in Chapter 2.
2 Late-Caledonian Orogeny. Post-Lower Devonian (Lower Old Red Sandstone), pre-Devono-Carboniferous (Upper Old Red Sandstone) movements, marked by a long period of non-deposition.
3 Hercynian Orogeny. Intra-Carboniferous and late-Carboniferous/Permian movements.
The principal structures present in the district are shown on Figure 12.

LATE-CALEDONIAN OROGENY

The final phases of the Caledonian Orogeny caused the uplift and erosion of the rocks of the district. Devono-Carboniferous strata rest with marked unconformity on the Lower Devonian rocks, there being no evidence of deposition during the Middle Devonian.

The late-Caledonian movements reactivated the large NE-trending Lammermuir and Dunbar–Gifford faults which form part of the Southern Upland Fault-belt (Anderson, 1951, pp. 93–98). The Lammermuir Fault separates Devono-Carboniferous rocks to the north-west from tightly folded Lower Palaeozoic strata and the Lower Devonian conglomerate to the south-east.

HERCYNIAN OROGENY

The faulting and folding affecting Carboniferous strata are generally ascribed to the Hercynian Orogeny. In the district, the detailed structures are not well known to the north-west of the Lammermuir Fault where exposure is poor and drift-cover obscures much of the rock. The Innerwick Fault separates the economically important area of Lower Limestone Group beds south-east of Dunbar from the older Great Conglomerate and Devono-Carboniferous strata.

Area north-west of the Lammermuir Fault

Detailed information in this area is sparse. The oldest strata are of Devono-Carboniferous age and are confined to a faulted belt between the Lammermuir and Dunbar–Gifford faults. They are poorly exposed except on the coastal section.

The Dunbar–Gifford Fault trends NE–SW, trending almost parallel to, and about 2 km north-west of the Lammermuir Fault. The position of the former fault is difficult to place accurately but is well marked on the coast just west of the Parade Vent at Dunbar. Strata between these major faults, of Devono-Carboniferous age, are seen to dip at be-

tween 10° and 25° to the south-east except near the fault planes where dips of up to 70° occur. Dips in inland exposures are somewhat haphazardly orientated but generally they are to the south-east and at about 10° to 20°. Exposures on the coast, east from Dunbar, show the strata dipping consistently east-south-east at dips ranging from 10° to 65°–70° near to the Innerwick Fault. Considerable disorientation of these rocks has been caused by the intrusion of volcanic necks, and faulting has caused local steepening of the dips.

To the north-west of the Dunbar–Gifford Fault, strata of Calciferous Sandstone Measures age, including the Garleton Hills Volcanic Rocks, are poorly exposed inland and are best known from coastal exposures. These strata are cut by a few NE-trending faults, the most notable one being the Gleghornie Fault (Figure 12). The rocks are also gently folded with the Whitekirk and Prestonkirk synclines preserving the lower flows of the Garleton Hills Volcanic Rocks, and the underlying sediments being brought to the surface by the Balgone, Crauchie and Traprain anticlines. These gentle folds all show south-west-plunging axes.

Lammermuir Fault

This fault is probably the most important structural element in the district and is thought to be part of the Southern Upland Fault-belt. It separates Lower Palaeozoic, Lower Devonian and Devono-Carboniferous in the south-east from Devono-Carboniferous rocks to the north-west. The position of the fault is well defined along most of its length but near Chesterhall, the NW-trending Innerwick Fault probably truncates the Lammermuir Fault. It is possible that the NE-trending faults at Lawrie's Den and Millstone Neuk may be continuations of the Lammermuir Fault. The downthrow across the Lammermuir Fault is to the north-west and may well be up to 300 m in places.

Area south-east of the Lammermuir Fault

Post-Silurian strata to the south-east of the Lammermuir Fault form the broad tract of the Great Conglomerate and Carboniferous strata preserved by the Innerwick Fault to the east of the conglomerate. In the Great Conglomerate, poor stratigraphical control seriously inhibits any structural interpretation. In general, dips, mainly recorded in the sandy intercalations, are gentle, averaging 10°–15° in a dominantly north-easterly direction. Fault planes are generally orientated NE–SW and NW–SE, but the displacement across these fractures is impossible to determine. The possibility that these are tension-fractures is discussed below.

In the Oldhamstocks area, Devono-Carboniferous strata dip gently to the east with the dips becoming steeper near the faults.

Calciferous Sandstone Measures strata between the Cove and Torness faults (Figure 12) are poorly exposed but

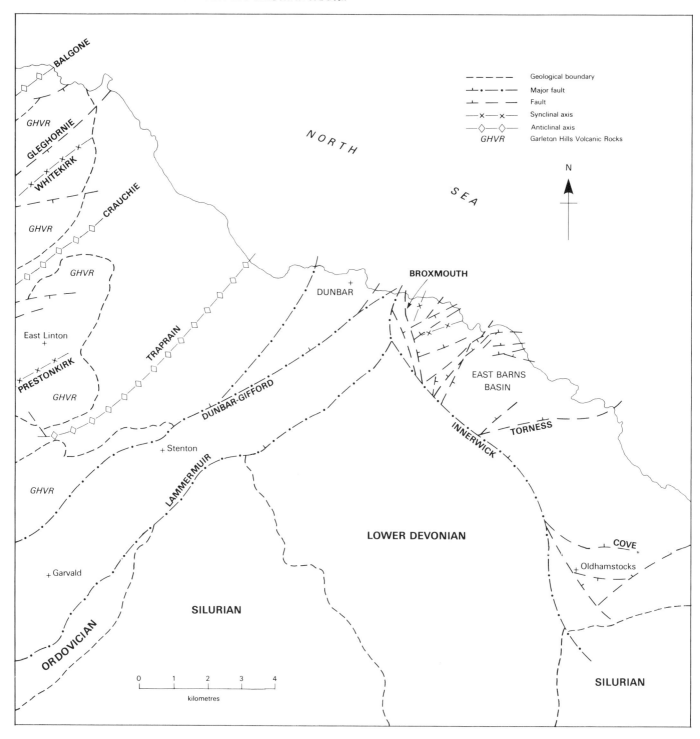

Figure 12 Diagram of main faults and folds in post-Silurian rocks

generally appear to dip gently to the east-north-east. In the Dunglass Burn [7588 7118] steep dips of up to 60° indicate the position of the Cove Fault, which is thought to have a displacement of over 80 m to the north which increases westwards towards the junction with the Innerwick Fault. This latter fault preserves Carboniferous strata downthrown to the east, the effective downthrow increasing northwards to a maximum of over 500 m near Easter Pinkerton [705 756]. The displacement decreases northwards, and two other

faults split off the Innerwick Fault (Figure 12). The most easterly fracture, the Broxmouth Fault, can be positioned accurately in the railway cutting [6974 7675], and on the coast at Fluke Dub [6938 7741], giving the fault a trend slightly west of north. At both localities adjacent strata are vertical or slightly overturned. In the railway cutting, a split in the fault preserves vertical beds of tuff, to the west of which steep dips of 75° to the west are seen in mudstones with cementstone bands. West of these beds an abrupt

Plate 14   Gentle folding in Upper Longcraig Limestone, Longcraig (D 1141)

change in lithology indicates the position of the central fault. The strata between this fault and the Innerwick Fault are mainly thin bedded, flaggy and, in parts, massive sandstones, with a general N–S strike and with steep or vertical dips.

At Fluke Dub, the beds west of the Broxmouth Fault are vertical tuffs and basalts succeeded by mudstones, sandstones and cementstone bands dipping westwards at up to 70°. These strata are cut off sharply by the Innerwick Fault and subsidiary fractures from beds of Devono-Carboniferous age dipping to the east-south-east. Crushing and steep to vertical dips are also seen at other localities along the Innerwick Fault south-east towards Oldhamstocks. In the Ogle Burn [7326 7222] and the Braidwood Burn [7285 7330] shattered greywacke occurs in the fault crush-zone and the Calciferous Sandstone Measures strata have steep dips immediately east of the crush-zone. The position of the Innerwick Fault south of the Ogle Burn is somewhat conjectural.

The structure of the Carboniferous Lower Limestone Group strata bounded by the Innerwick and Torness faults is best described in two parts. The southern half is a broad, open basin, here named the East Barns Basin (Figure 12), the axis of which plunges gently north-eastwards. The northern half of the fault-bounded area is much affected by NE-trending faults which, in association with some minor synclinal folds, have caused the repeated outcropping of the principal limestones.

The southern part shows generally gentle dips, averaging about 10° to the east, from Catcraig to Barns Ness. Southwards from the lighthouse the direction of dip gradually rotates to north-east, and from Skateraw Harbour to Torness Point it becomes northerly to slightly west of north. Some gentle rucking with a dominant axial trend 15° west of north, is particularly well seen in the Middle Skateraw Limestone from Skateraw Harbour to Torness Point. These gentle folds all plunge northwards at low angles.   AD

Tectonic control of basaltic dykes in the Great Conglomerate

The Great Conglomerate of the district is younger than the late-Caledonian porphyrite and porphyry dykes which are never seen to cut it. Pebbles of the dyke rocks have been recorded in the conglomerate by Irving (1930, pp. 538–539). The conglomerate is cut by a few dykes of the Stephanian quartz-dolerite suite which also cut the Silurian rocks. The most remarkable feature, however, is the presence of a suite of basaltic dykes thought to be of Carboniferous age which are restricted to the conglomerate but whose trend is oblique to that of the belt of conglomerate. It

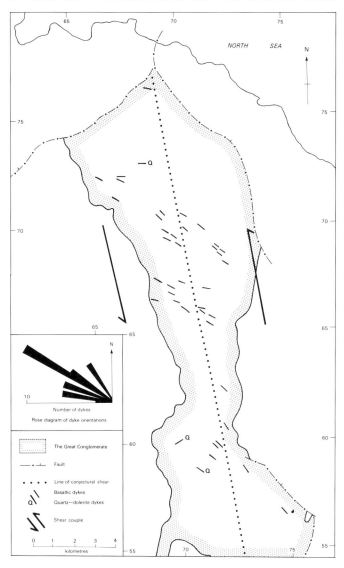

Figure 13 Diagram of basic dykes cutting Great Conglomerate and orientation of possible shear-couple related to their pattern of intrusion

seems most likely that they are intruded into tension fractures on either side of a shear fault with sinistral sense of movement. Cloos (1955, pl. 3) produced this effect experimentally with development of *en echelon* tension-fractures at about 45° to the shear-couple and the production of open, short gashes only within the flexure zone. Such a situation could develop without the actual production of a major dislocation or fault. The orientation of a shear-couple related to the emplacement of the dykes within tension fractures is shown in Figure 13.

It is probable that the deep valley in which the extant deposit of the Great Conglomerate was deposited was eroded along a fault. Reactivation of this crustal fracture involving transcurrent movement subsequent to the deposition of the conglomerate would account for the attitude of the dykes and their restriction to a zone along the axis of the pre-Great Conglomerate valley.   RWE

CHAPTER 10

Quaternary

INTRODUCTION

There is no evidence of sediments having been deposited in the district during the long interval between the close of Lower Carboniferous times and the Quaternary. Any strata laid down during this period have been removed by subsequent erosion.

The drainage-pattern of eastern Scotland is thought to have been initiated on an easterly-tilted, marine-planed surface formed during the Cretaceous. The east-flowing River Forth and the River Tweed were initiated as consequent drainage on this slope and the River Tyne and the White-adder Water are the main tributaries in the district. Erosion during the Tertiary cut down into the underlying rocks and, as hard or soft rocks were encountered, the simple river pattern became more complex (Bremner, 1942; Linton, 1951; Sissons, 1967).

Glacial deposits have obscured and infilled the valleys of the pre-Glacial river system in the Lothians, related to a sea-level lying well below the present one. Although the courses of several buried channels are well known in areas to the west (Mitchell and Mykura, 1962, fig. 25) there are only two places in the district where there is evidence of buried channels. At Peffer Sands geophysical evidence has shown the existence of a deep, drift-filled channel, marking a former mouth of the River Tyne which previously took a more northerly route through the Garleton Hills (McAdam and Tulloch, 1985). The present estuary of the River Tyne may lie on a similar buried channel at the former mouth of the Biel Water.

AGE OF THE GLACIATION AND QUATERNARY DEPOSITION

During the Pleistocene period there were several glacial and inter-glacial stages. Although these have been recognised in continental Europe and to some extent in England, in central Scotland the latest glaciation (Devensian) obliterated evidence from all earlier glaciations and interglacial periods. The Devensian glaciation, equated with the Würm stage in the Alps and the Weichsel stage in North Germany (Flint, 1957, pp. 404–413), lasted from 80 000 to 10 000 years BP.

Warming of the climate and retreat of the Devensian ice-sheet began about 18 000 BP and the district may have been ice-free by 13 000 years BP. Readvance of the ice and cooling of the climate occurred between 11 000 and 10 300 years BP. During the late-Glacial period, when the ice-sheet was melting, most of the glacial drainage channels were cut and the stratified glacial deposits of sand and gravel or lacustrine silt and clay were laid down, and the high late-Glacial raised beaches were formed.

The main Flandrian deposits comprise the post-Glacial raised beach and associated blown sand deposits, which formed during the Flandrian Marine Transgression between 7000 and 5000 years BP. Other post-Glacial deposits include peat, river-terrace alluvium, flood-plain alluvium and lake deposits. Recent deposition is restricted to coastal beach deposits, sand dunes and river flood-plains.

DIRECTION OF ICE-FLOW

The regional ice-flow pattern in central Scotland was investigated last century using evidence from lineations and indicator stones. Snow and ice accumulating in the Southern Uplands around Broad Law, Hartfell and White Comb flowed north-eastwards, encountering south-easterly flowing Highland ice from the Loch Lomond area. The merged ice flowed almost due eastwards across the Lothians, being deflected east-north-east by the mass of the Lammermuir Hills, before turning at Dunbar to the south-east along the coast. In the south-east of the district, Southern Upland ice was diverted south-eastwards along the Whiteadder valley.

Passage of the ice-sheet moulded the underlying rocks and the ground moraine or till which it deposited. Ice, being a selective erosive agent, left the hard igneous rocks as upstanding crags and cut deep hollows into the softer sedimentary rocks. Ice-moulded features are particularly prominent in the north part of East Lothian (Figure 14). Glacially striated rock surfaces are, however, uncommon in the district. One or two have been recorded in the craggy lava terrain of Whitekirk Hill and north-east of Traprain Law and also along the coast. Crag and tail features are common in the volcanic hills in the north-west of the district. Spectacular tails are associated with Traprain Law and Knockenhair at Dunbar. Tail features also occur in the lee of volcanic plateaus at Whitekirk, Tyninghame and East Linton. Lineated boulder clay has formed, away from the effect of crags, in the ground between the River Tyne and the Biel Water, in some cases extending for over 3 km, although later glacial drainage has interrupted the lineations in places. Lineations give a more consistent, more accurate indication of flow-direction than glacial striae, as the basal ice is easily deflected by irregularities in the rock surface (McAdam and Tulloch, 1985, table 4). These show the direction of ice flow to have been east-north-east along the western edge of the district, becoming due east at Dunbar.

TILL (BOULDER CLAY)

Till covers much of the low-lying areas of Upper Palaeozoic rocks north of the Lammermuir Fault and in the Dunbar area. It is also extensive on Lower Palaeozoic rocks in the low-lying areas of the Tweed basin in the south-east of the district. In the upland areas till occurs only in some valleys, whereas in most of the low-lying areas, it is commonly 5 to

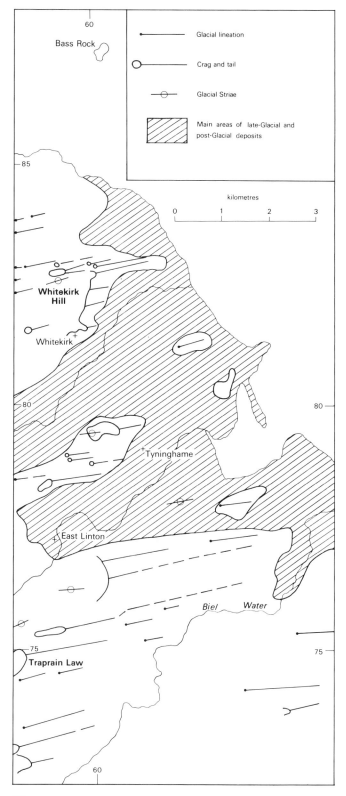

Figure 14 Ice-moulding in the north-western part of the district

10 m thick. In areas of volcanic rocks the till is thin and patchy. Thick accumulations may occur in buried channels at Peffer Sands and Tyne Sands.

As most of the constituent material is locally derived, the colour, stone-content and matrix of the till correspond closely to the underlying rocks. Erratic stones such as Highland schists, though significant in understanding ice-flow directions, are present in only small proportions. In the north-west of the district the till is commonly a mottled grey, purple and orange, silty or sandy clay with pebbles of basalt, trachyte, dolerite and other igneous rocks mixed with sandstone, limestone and other sedimentary rocks. In areas of Devono-Carboniferous rocks the till is a red or purple, sandy clay with dominantly red sandstone pebbles. Greywacke and quartzite cobbles are marked in till derived from Lower Devonian conglomerates. In areas of Lower Palaeozoic rocks the till is brown and sandy with fragments almost entirely of greywacke, siltstone and shale. Weathering has commonly altered the top one or two metres of the till to a brown, less consolidated deposit.

DEGLACIATION

As the climate improved and the ice-sheet covering the district began to melt and break up, Highland ice retreated northwards towards the Firth of Forth and Southern Upland ice retreated south to form an isolated ice-cap with glaciers in valleys such as the Whiteadder. Ice-free ground appeared along the slopes of the Lammermuir Hills, and increased in area as the ice-margin gradually receded to successively lower levels. The large volumes of glacial meltwater released by the melting ice cut numerous glacial meltwater channels and laid down stratified glacial deposits, mainly of sand and gravel, along the foothills, along valley sides and on the coastal plain (Figure 15). Ice lying in the low ground blocked most of the river valleys, damming the meltwater, which was forced to cut new channels along the ice-margin and across cols into other valleys. Channels vary greatly in length and depth, from short channels only a metre or two deep, to channels several kilometres long and up to 30 m deep. Many channels were abandoned and are now dry valleys, while others are still utilised by the modern drainage. The glacial drainage flowed generally north-eastwards, parallel to the slopes of the Lammermuir Hills, and down the gradient of the ice-sheet. The majority of channels are contour channels cut by meltwater flowing between the ice-margin and the hills, while some meltwater escaped under the ice to lower ground down glacial chutes, perpendicular to the contours.

DETAILS

The numbers in round brackets in the following section refer to Figure 15 which covers the northern part of the district.

Whiteadder valley

Most of the high-level channels are cut into rock, whereas the few channels in lower ground are cut into till. A succession of these features cut across the rock ridge from Langton Edge to Knock Hill [74 55]. Marginal channels were cut on the slopes east of the Watch Water Reservoir [66 56] and on Dirrington Hill [68 55]. Networks of downhill channels have been recognised in the upper valley of the Whiteadder Water at Cranshaws [62 68], Harehead [69 63], Gowl Burn [68 64], Crystal Rig [66 66] and Hungry Snout [66 63]. Many

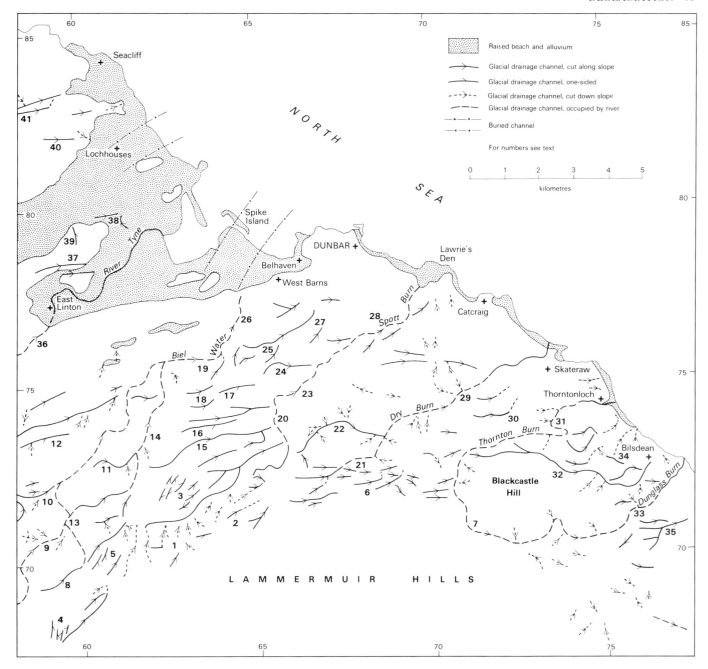

Figure 15 Map of glacial drainage

of the main valleys in the basin carried meltwater, but it is not clear how much of the erosional or depositional features can be attributed to this agency.

Lammermuir Hills

Virtually no meltwater features are present on the gentle south slopes of the Lammermuir Hills whereas numerous such features have been cut on the steep north slopes. Marginal channels occur at Deuchrie Edge (1), Lothian Edge (2) and both sides of Deuchrie Dod (3), the last described in detail by Sissons (1975). Similar marginal channels were cut in hills of till at Star Wood (4) and east of Thorter Reservoir (5). Meltwater found an escape route through the Lammermuir Hills by Hartside and Woodhall, carving out deep channels (6) and depositing high ridges of greywacke gravel. Much

of this meltwater found its way into the spectacular alluvium-floored channel round the south of Cocklaw Hill (7) contributing to the coastal spreads of sand and gravel south and east of Oldhamstocks.

Lammermuir foothills

Standstill of the ice-margin at various levels along the foothill plain led to the production of sinuous, glacial-meltwater channels cut deep into rock and the deposition of extensive spreads of sand and gravel, in the form of dissected kame-terraces, ridges and mounds. In the west, a series of channels (8 to 12) mark successively lower levels of the ice-margin, the meltwater being collected and carried northwards by the deep channels (13, 14) now occupied by the Thorter Burn and Sauchet Water. A similar set of channels (15 to 19) farther east lead into the Spott Burn (20). Still further east several channels (21 to 25) also mark levels of the ice-margin. There is, however, a lack of glacial-lake strand-lines.

Coastal Plain

At a later stage several channels (26 to 33) carried the meltwater on to the coastal plain depositing long spreads of sand and gravel from Dunbar to Oldhamstocks and beyond the boundary of the district. Within these deposits networks of channels (34, 35) further dissect the sand and gravel.

North of River Tyne

Part of the Tyne Gorge (36), cut by meltwater, lies in the district. The meltwater issuing from the gorge carried large amounts of material which formed the extensive raised beach and glacial-marine deposits bordering the Tyne valley. Other channels cut in this environment were formed at Preston (37) and Tyninghame (38). Ice banked up against the lava hills cut channels at Lawhead (39) and Whitekirk (40). Meltwater from the west cut channels at Gleghornie (41).

GLACIAL SAND AND GRAVEL

Stratified deposits of sand and gravel were laid down by meltwater as the ice-sheet melted and the ice-front retreated towards the Firth of Forth. Such deposits, laid down close to the ice-front or standing ice, took the form of kames, kettles, mounds and esker-ridges. Fluvioglacial deposits are mainly valley terraces, formed away from ice margins. Several stages of glacial retreat occurred, each with glacial and fluvioglacial deposits, the later episodes dissecting or adding to earlier deposits, blurring the distinction between glacial and fluvioglacial deposits.

The main deposits occur along the foothills from Snawdon to Garvald, through the Hartside–Woodhall gap, down the Biel Water at Stenton, along the River Tyne at East Linton, in a coastal spread from West Barns to Dunglass, and inland around Oldhamstocks.

Rapid variation in grain-size and lithological content occur throughout the deposits (McAdam, 1978). The high ridges along the Lammermuir foothills at Stoneypath and Oldhamstocks and through the Hartside–Woodhall gap are commonly coarse boulder gravel, with little sand. Pebbles in the gravel are mainly of greywacke, with the remainder composed of red sandstone. In the lower ground, the deposits have more varied forms, contain sand and fine-grained gravel, with the proportion of locally derived Carboniferous rocks increasing rapidly northwards and eastwards. In the Tyne valley, the deposits are in the form of glacial mounds and kames. At East Linton they have been modified by the action of the late-Glacial sea. The bulk of the finer sediment, silt and clay, was washed out into this late-Glacial sea, contributing to the raised beaches.

CHANGES IN SEA-LEVEL, LATE-GLACIAL TO PRESENT

Sea-levels have varied considerably since the start of the last deglaciation (Donner, 1963; Sissons, 1967; Goodlet, 1970).

The variations were complex and caused by two major effects. Firstly, during glaciation world sea-level was lower due to the locking up of water in ice caps. Secondly, isostatic depression of the land, caused by the weight of the ice, was followed by uplift when the ice disappeared. Because isostatic uplift was generally predominant, sea-levels have been mainly higher than at present resulting in raised-beach features and deposits. Also, because the weight of the ice was greatest over the centre of Scotland, the subsequent isostatic uplift was also greatest there and the older beaches tilt away from the centre (Donner, 1963).

The earliest beaches in the district are of late-Glacial age, formed from about 14 000 to 11 000 years BP when sea-level was at 20 to 25 m OD. Deposits of this age occur in the valleys of the Pilmuir Burn, Peffer Burn and the River Tyne. From about 12 000 to 10 000 years BP evidence from other areas indicates a sea-level well below the present one. From about 7000 to 5000 years BP sea-level was generally at about 8 m OD and formed the narrow but persistent post-Glacial raised beach along much of the coast, before falling gradually to its present level.

LATE-GLACIAL RAISED BEACHES

DETAILS

Pilmuir Burn

The deposits in this small valley (Figure 16) are equivalent to those in the Peffer valley, are at the same level as them, and were laid down at the same time. Estuarine clay covers the ground from 22 m down to 8 m OD where it is covered by the later post-Glacial raised beach. The deposit lies on till. The back level is generally at about 20 m OD on the north bank, but the deposit only reaches 8 m OD on the south bank possibly because of later erosion. Terraced gravel deposits mainly on the north of the valley are found at levels up to 33 m OD. The gravel, 1.5 m thick, lies on 3 m of alluvial clay in a cliff section at Chapel Brae [6145 8388].

Peffer Burn (east)

Estuarine deposits cover the bottom of the valley of the Peffer Burn in an area trending NE–SW, 7 km long by 1.5 to 3 km wide. The deposits lie entirely below 22 m OD, are banked up against till and glacial sand and gravel, and are overlain below 8 m OD by the post-Glacial raised beach. Much of the deposit is silt and clay which blankets the pre-existing topography without forming any flats or shorelines. At Scoughall the upper limit against till falls to 8 m OD, and two inliers of till occur at Brownrigg [620 813] and Tyninghame Links [628 804]; marine deposits are absent here as a result of later erosion or because of protection by stagnant ice. The beach deposit is characterised by small side channels 1 to 6 m deep flowing towards the Peffer Burn, usually cut entirely within the area of the deposit, though some start higher up the valley side. The deposit is rarely exposed in natural sections. It forms a grey or yellowish stoneless clay soil, in places becoming silty or finely sandy. In the higher ground towards Tyninghame, sands and fine gravels have been reworked from the glacial sands and gravels. A traverse of shallow proline auger holes, drilled from New Mains [605 825] to Tyninghame House [620 800], proved deposits, at least 6 m thick, of stoneless, grey, blue and purple clay, with beds of sand and fine gravel.

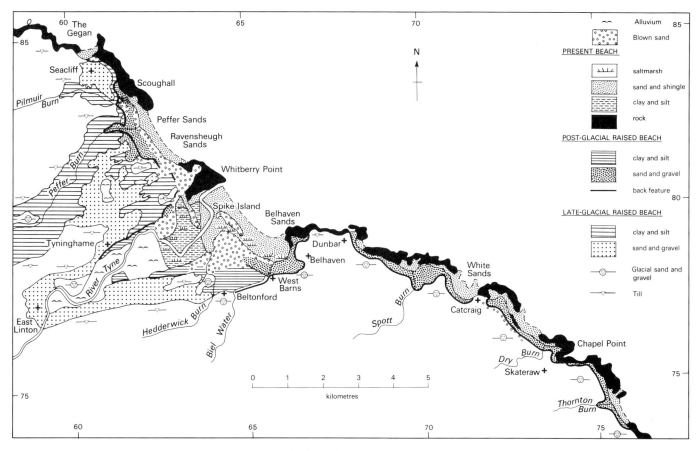

Figure 16 Map of raised beach and present beach deposits

River Tyne

Late-Glacial estuarine deposits occur round the estuary of the River Tyne, lying below 30 m OD. They include modified glacial sand and gravel deposits. At the mouth of the Tyne Gorge, the source of most of the material, the deposits occur as high as 30 m OD. The back feature of the deposits falls to about 22 m OD at Beltonford [640 776] and Tyninghame. At levels below 8 m OD, the deposits are overlain by post-Glacial raised beach deposits and by the alluvium of the River Tyne. Kame and kettle sand and gravel deposits occur within the beach deposits at Kirkland Hill [618 778] and ridged gravel deposits occur south of Preston Mains [600 780]. An inlier of till is present at Tynefield [632 782]. Terrace features are well developed along the south side of the valley from Phantassie [595 771] to Beltonford, and at several levels around Tyninghame. Evidence of the lithology is derived almost entirely from soils. The deposits consist mainly of fine- to medium-grained gravel and sand with local areas of clay. A road cutting at Beltonford exposed a section of red gravel (1.5 m), on bedded sand (1.5 m), on laminated clay (2.0 m).

POST-GLACIAL RAISED BEACHES

DETAILS

The Gegan to Ravensheugh Sands

Along this part of the coast, the post-Glacial raised beach forms a narrow bench backed by a prominent, steep slope with the back feature at 8 m OD. Small patches of beach occur below Seacliff [608 845]. A beach, 100 m wide at Scoughall in the north, stretches for 3 km to Lochhouses Links [623 817] in the south where it widens to 1 km. The deposits consist of shell-sand and shingle, as seen in soils and coastal sections, and proved up to 3.5 m thick in auger holes. Shingle storm-ridges cut off two small inlets forming small lagoons near Lochhouses where alluvium has been deposited. Ridges of blown sand lie along the seaward edge of most of the deposit and inland there are isolated mounds. In recent years the raised beach of Lochhouses Links has been uncovered by extensive quarrying of the blown sand.

Tyne Estuary

During the time when sea-level was at 8 m OD, marine deposits filled the lower part of the Tyne Estuary in the same way as is occurring at the present day. These older deposits have been greatly dissected by more recent river erosion, but nevertheless, show similarities to the present deposits. Upstream, red, silty clay alluvial deposits in sloping terraces merge gradually seawards into yellow, silty clay estuarine deposits in flat terraces, with a sharp change into longshore sand deposits. The estuarine silty clay forms broken narrow terraces north and south of the present alluvium and also forms a wide, flat, raised beach north of Beltonford and West Barns. Sandy deposits occur north-east of Tyninghame House [628 802], but they are mainly covered by blown sand, all but obscuring the back feature. On the south side of the river the raised-beach sand underlies an area from Spike Island [645 795] to Belhaven [660 785]. Sections exposed along the over-deepened valley of the Hedderwick Burn [6397 8363] show the following sequence:

	Thickness m
Blown sand	0.3
Grey clay	0.3
Bedded sand and silt	1.8
Dark clay	0.9

and along the south side of the estuary where recent erosion has exposed the section [6355 8366]:

	Thickness m
Blown sand	1.8
Grey clay	0.1
Bedded sand and silt	1.8
Obscured	1.2
Dark grey clay with plant fragments	0.3

These sections occur near the junction between estuarine and longshore deposits. Sections of at least 1.2 m of bedded, shelly sand in the longshore deposits are also exposed by erosion along the shore south of Spike Island.

Dunbar to Thorntonloch

At Dunbar, Catcraig, Skateraw and along the coast south-east of Bilsdean, the post-Glacial raised beach is absent where high cliffs fall directly to the foreshore. Elsewhere, long narrow strips of this beach fringe the coastline and are backed by prominent cliff back-features. Normally the raised beaches are only 100 to 200 m wide. An extensive area of raised beach north-west of Broxmouth has recently (1985) been destroyed and obscured during limestone quarrying operations. At Lawries Den [705 784] post-Glacial raised-beach platforms preserved in the intertidal zone have been dissected by recent marine erosion. Deposits of the raised beach, exposed in places along H.W.M., consist mainly of shell-sand and shingle with shelly faunas associated with rocky, high-energy environments.

PEAT

Immediately following glaciation, conditions encouraged extensive colonisation by birch-hazel-pine and heather-sphagnum floras. From these, hill peat developed on till and rock surfaces in the Lammermuir Hills. In the lower ground, bog-floras formed basin-peat associated with alluvial deposits. Only remnants of peat have survived subsequent erosion and human exploitation (Ragg and Futty, 1967).

FRESHWATER ALLUVIUM

Most of the modern river valleys as well as many of the dry, glacial drainage channels are floored by alluvial deposits.

Plate 15   Glacial overflow channel, now dry valley, Braidwood (D 1137)

Plate 16 Shingle of post-Glacial raised beach, resting on bedded sandstones, Skateraw (D 1138)

Generally these are narrow strips of flood-plain alluvium which consist of interbedded gravels, sands, silts and clays. These constituents occur in constantly varying proportions though normally the gravels and coarser sands are commoner in the steeper headwaters and higher terraces, whereas the finer deposits are predominant in the gentler lower reaches and flood-plain terraces. Lake deposits of laminated silt and clay fill hollows in the glaciated till surface.

Where the River Tyne enters the district, it flows in a rock gorge cut in basaltic lavas, with little or no alluvium. At East Linton the valley opens out and the river runs for 6 km in a wide alluvial plain which merges into estuarine deposits. The alluvium forms sloping dissected terraces at three main levels. The lowest deposit, of fine sand and silt, is the present flood-plain which is confined to narrow strips along the river. The main spread of alluvium forms terraces at 2 to 4 m above the river and consists of reddish sand and silt upstream, and reddish or yellowish silty clay downstream. The highest fragmented terraces, of silt, sand and gravel, lie 6 to 8 m above the river and fall from 15 m OD at the gorge mouth to the post-Glacial raised beach level at 8 m OD.

Where cut in rock, the River Whiteadder and its many tributaries form narrow valleys with little alluvium but where excavated in till the valleys are open with wide alluvial plains. Complex terraces are related to specific nick-points reflecting various episodes of down-cutting.

BLOWN SAND

Dunes of blown sand have developed along the coast wherever sandy, post-Glacial raised beach and present beach deposits are present, particularly around the Tyne estuary. The deposits are still being formed though some may date back to post-Glacial times. Dune ridges, 6 to 8 m high, form barriers along High Water Mark as at Peffer Sands, Belhaven Sands and on Spike Island. More widespread deposits of low irregular dunes occur on the raised beach, as at Lochhouses Links, Links Wood and West Barns Links. The deposit is a fine-grained, cross-bedded sand, consisting mainly of quartz with about 10 per cent shell debris and 5 per cent ferro-magnesian minerals. Blown sand has been worked in small pits at Lochhouses Links [627 817] and West Barns Links [643 788] (McAdam, 1978).   ADM

PALAEONTOLOGY

The only fossiliferous late-Glacial deposit recorded from the district was found at West Barns [659 784], where a blue-grey, laminated clay, formerly worked for brick-making but no longer exposed, included a band with numerous specimens of the cold water starfish *Ophiolepis gracilis* Allman (Allman, 1863, pp. 101–104). This fossil is present in the

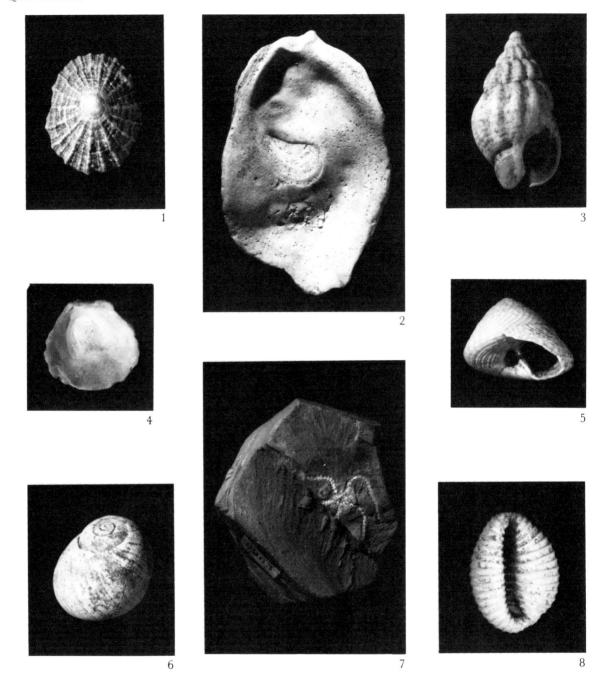

Plate 17 Quaternary fossils

1, 3 and **6** Molluscs from the raised beach on the east side of
Belhaven Bay [663 790]

2, 4, 5 and **8** Molluscs from the raised beach on the south side
of Chapel Point [739 758]

7 Starfish from late-Glacial clay at Dunbar brickworks, West
Barns [659 784]

1 *Patella vulgata* × 1.5
2 *Ostrea edulis* × 1
3 *Nassarius incrassatus* × 5
4 *Monia patelliformis* × 1
5 *Gibbula cineraria* × 2
6 *Littorina littoralis* × 2
7 *Ophiolepis gracilis* × 1.5
8 *Trivia sp.* × 3

arctic assemblage of the Errol Beds of the St Andrews and Montrose areas, with which the fauna from West Barns probably correlates. *O. gracilis* is a delicate species, and the abundance of well-preserved specimens occurring suggests tranquil depositional conditions.

The post-Glacial raised beach is exposed at a number of locations along the shore between Seacliff and Bilsdean and examples of fossils from two localities, Belhaven Bay [663 790] and Chapel Point [739 758], are shown in Plate 17. The raised-beach assemblages do not appear to differ significantly from those of the present beaches. In areas where numerous rock outcrops occur, the raised beach faunas are generally dominated by rock dwelling gastropods such as *Littorina spp.*, *Nassarius incrassatus* (Ström), and *Patella vulgata* (Linné).

Faunas recovered from the blown sand deposits in the Hedderwick Burn and on the shore on either side of its mouth [640 788] are similar to those found on the present beach, and the absence of rock outcrops in the area is reflected in the scarcity of rock dwelling molluscs. The most abundant gastropods are *Onoba semicostata* (Montagu), and *Rissoa parva interrupta* (Adams), whilst the bivalve population is dominated by *Spisula spp.*, *Venus spp.*, and anomiids.

Near to Strand House, Broxmouth [707 773], the fauna from a post-Glacial marl was found (Young *in* Howell and others, 1866, p. 67) to contain freshwater molluscs. These include species of *Lymnaea* and *Planorbis*, together with some terrestrial elements.   DKG

CHAPTER 11

Economic geology

LIMESTONE

Economic deposits of limestone are confined to the area south-east of Dunbar. The limestones are of Lower Limestone Group age, and were initially exploited from many small quarries for the production of agricultural lime and flux for iron smelting. All these quarries and their attendant kilns, notably at Oxwell Mains, Catcraig and Skateraw harbour, are now abandoned. Some small-scale quarrying was also carried out in the intertidal zone at Catcraig, Chapel Point and Skateraw harbour.

The thickest, and most widely exploited, limestones are the Middle Skateraw and Upper Longcraig limestones. Extensive quarrying of these two beds is carried out at present at Dunbar Quarry (Plate 18) for the manufacture of cement. The available and usable resources of these limestones are well proven, and are owned by the present operators.

ROADSTONE AND AGGREGATE

No roadstone quarries are at present being worked in the area. Potential sources of road metal and aggregate, although of variable quality, occur in the eastern end of the Garleton Hills Volcanic Rocks outcrop. Rock of more uniform quality occurs in the trachyte intrusion at Garvald, the granite of Cockburn Law and the dolerite sill at Harelawcraigs Plantation at the extreme south-east corner of the area. Quarrying was formerly carried out at Traprain Law, and on a very restricted scale at Millstone Neuk on the coast south-east of Dunbar.

Vast resources of Silurian greywacke and shale are present. These sediments are variable, composed of alternating bands of greywacke and shale. The greywackes in some parts occur as fairly thick bands which form valuable sources of hard, resistant rock with high polished-stone values. Ag-

Plate 18 Dunbar Quarry, showing removal of glacial overburden, Middle Skateraw Limestone in upper third of face and Upper Longcraig Limestone in lower third (D 1142)

gregate from these greywackes may, however, be troublesome in some applications due to shrinkage. Old abandoned small quarries, mainly for stone for drywalling, are very common over the area of outcrop.

BUILDING STONE

Red-brown sandstone of Devono-Carboniferous age was formerly quarried at Broomhouse Quarry 2.5 km south of Dunbar. Here some 9 m of massive, cross-bedded, medium-grained sandstone, with coarse and pebbly bands, was worked for use as a local building stone. Reserves of sand-stones of similar age are extensive, but of variable quality.

COAL

Coal mining took place on a restricted scale during the last century in the vicinity of Lawfield some 10 km south-east of Dunbar. The coal worked ranged up to a maximum thickness of 0.5 m. The coal is of Calciferous Sandstone Measures age and was formerly exposed on the foreshore [7605 7316] where about 0.25 m of poor coal was seen. Boreholes put down in 1907 proved an old waste, but the extent of the working is unknown. There are no economic reserves of coal in the Dunbar district.

SAND AND GRAVEL

Glacial meltwater deposits occur along the margins of the Lammermuir Hills, and along a coastal belt from Dunbar to south of Oldhamstocks. Other areas of sand and gravel occur in the vicinity of Halls [654 727] and from Woodhall [685 726] to Thurston Mains [710 730]. The deposits do not occur in economic quantities over all these areas but those that can be classed as economic resources were detailed by McAdam (1978).

The deposits immediately adjacent to the Lammermuir Hills generally form ridges and mounds of dominantly greywacke gravel. On the lower ground terraced deposits occur containing clasts of a more mixed provenance, including sandstone and igneous pebbles as well as greywacke. Some of the ridge deposits are up to 15 m thick, but generally deposits are much thinner with an average thickness of from 3 to 5 m. At present these deposits are not exploited on any significant scale.

BLOWN SAND

At Lochhouses Links blown sand is worked on a small scale for use as building sand. Dunes and mounds of blown sand, forming a potential resource, occur southwards from Peffer Sands to Belhaven Bay.   AD

WATER SUPPLY (GROUNDWATER)

Groundwater is an important economic resource in the Devono-Carboniferous and Carboniferous strata. A Devono-Carboniferous sandstone at Belhaven supplies a brewery from a single 80-m deep borehole which has a sustained yield of 10 ls^{-1} (8000 gallons per hour). Two slightly deeper boreholes penetrate the Lower Limestone Group and the underlying Calciferous Sandstone Measures at Dunbar Quarry. They have a combined capacity of 15 ls^{-1}. A 140-m deep borehole overflows at 10 ls^{-1} at nearby Catcraig Quarry. Recharge is available to the aquifers penetrated by these boreholes via the drift, wherever it is sufficiently permeable to allow ingress of water from the surface. Although formal evaluation of the aquifer resources, including those of the Calciferous Sandstone Measures which form an aquifer south-east of Dunbar, has not yet been made, considerable further groundwater development could be supported as the water table maintains a roughly constant average annual level.

Water quality is good. The water from the Devono-Carboniferous is moderately mineralised and has a high bicarbonate concentration (300 mg l^{-1}). A lower level of mineralisation with a bicarbonate concentration of about 170 mg l^{-1} typifies water from the Carboniferous.

Several private domestic supplies draw on other aquifers. In the Ordovician, Silurian and Lower Devonian, supplies are small because storage and transport of groundwater is limited to the secondary porosity created by cracks and joints. Small groundwater supplies are also available from coarse horizons in the drift, such as alluvium, raised beach and glacial meltwater deposits. Water quality from these aquifers is generally good.   NSR

REFERENCES

ALLMAN, G. J. 1863. On a new fossil *Ophiuridan*, from the post-Pliocene strata of the valley of the Forth. *Proc. R. Soc. Edinburgh*, Vol. 5, 101–104.

ANDERSON, E. M. 1951. *The dynamics of faulting and dyke formation with application to Britain* (2nd edition). (Edinburgh: Oliver and Boyd.)

BALSILLIE, D. 1936. Leucite-basanite in East Lothian. *Geol. Mag.*, Vol. 73, 16–19.

BERGSTRÖM, S. M. 1971. Conodont biostratigraphy of the Middle and Upper Ordovician of Europe and eastern North America. *Mem. Geol. Soc. Am.*, No. 127, 83–157.

BENNETT, J. A. E. 1945. Some occurrences of leucite in East Lothian. *Trans. Edinburgh Geol. Soc.*, Vol. 14, 34–52.

BREMNER, A. 1942. The origin of the Scottish river system. *Scott. Geogr. Mag.*, Vol. 58, 15–20, 54–59, 99–103.

CAMPBELL, R. and STENHOUSE, A. G. 1934. The occurrence of nepheline and fayalite in the phonolitic trachyte of the Bass Rock. *Trans. Edinburgh Geol. Soc.*, Vol. 13, 126–132.

CLARK, R. H. 1956. A petrological study of the Arthur's Seat Volcano. *Trans. R. Soc. Edinburgh*, Vol. 63, 37–70.

CLOOS, E. 1955. Experimental analysis of fracture patterns. *Bull. Geol. Soc. Am.*, Vol. 66, 241–256.

CLOUGH, C. T., BARROW, G., CRAMPTON, C. B., MAUFE, H. B., BAILEY, E. B. and ANDERSON, E. M. 1910. The geology of East Lothian. *Mem. Geol. Surv. G.B.*

CRAIG, G. Y. and WALTON, E. K. 1959. Sequence and structure in the Silurian rocks of Kirkcudbrightshire. *Geol. Mag.*, Vol. 96, 209–220.

— and DUFF, P. M. D. 1975. (Editors). *The geology of the Lothians and south east Scotland.* (Edinburgh: Edinburgh Geological Society, Scottish Academic Press.)

CRAMPTON, C. B. 1905. The limestones of Aberlady, Dunbar and St Monans. *Trans. Edinburgh Geol. Soc.*, Vol. 8, 374–378.

CURRIE, E. D. 1954. Scottish Carboniferous goniatites. *Trans. R. Soc. Edinburgh*, Vol. 62, 527–602.

DAY, T. C. 1928. The volcanic vents on the shore between North Berwick and Tantallon Castle. *Trans. Edinburgh Geol. Soc.*, Vol. 12, 41–52.

— 1930a. Volcanic vents on the coast from Tantallon Castle eastwards to Peffer Sands, and at Whitberry Point. *Trans. Edinburgh Geol. Soc.*, Vol. 12, 213–223.

— 1930b. The intrusive rock of Frances Craig, and the teschenite of Ravensheugh. *Trans. Edinburgh Geol. Soc.*, Vol. 12, 260–261.

DONNER, J. J. 1963. The Late- and Post-Glacial raised beaches in Scotland, 2. *Ann. Acad. Sci. Fenn. Sar. A3 Geol. Geogr.*, Vol. 68, 1–13.

FLETT, J. S. 1908. On the mugearites. *Mem. Geol. Surv. G.B. Summ. Prog. for 1907*, Appendix 1, 119–126.

FLINT, R. F. 1957. *Glacial and Pleistocene geology.* (New York: Wiley.)

FRANCIS, E. H. 1962. Volcanic neck emplacement and subsidence structures at Dunbar, south-east Scotland. *Trans. R. Soc. Edinburgh*, Vol. 65, 41–58.

— FORSYTH, I. H., READ, W. A. and ARMSTRONG, M. 1970. The geology of the Stirling district. *Mem. Geol. Surv. G.B.*

GEORGE, T. N., JOHNSON, G. A. L., MITCHELL, M., PRENTICE, J. E., RAMSBOTTOM, W. H. C., SEVASTOPULO, G. D. and WILSON, R. B. 1976. A correlation of Dinantian rocks in the British Isles. *Spec. Rep. Geol. Soc. London*, No. 7.

GOODLET, G. A. 1970. Sands and gravels of the southern counties in Scotland. *Rep. Inst. Geol. Sci.*, No. 70/4.

GREIG, D. C. *In press.* Geology of the Eyemouth district. *Mem. Geol. Surv. G.B.*

HARRIS, A. L. (Editor.) 1985. The nature and timing of orogenic activity in the Caledonian rocks of the British Isles. *Mem. Geol. Soc. London*, No. 9.

HOWELL, H. H., GEIKIE, A. and YOUNG, J. 1866. The geology of East Lothian. *Mem. Geol. Surv. G.B.*

IRVING, J. 1930. Four 'Felstone' intrusions in central Berwickshire. *Geol. Mag.*, Vol. 67, 529–541.

KELLING, G. 1961. The stratigraphy and structure of the Ordovician rocks of the Rhinns of Galloway. *Q. J. Geol. Soc. London*, Vol. 117, 37–75.

LAGIOS, E. and HIPKIN, R. G. 1979. The Tweeddale Granite—a newly discovered batholith in the Southern Uplands. *Nature, London*, Vol. 280, 672–675.

LAMONT, A. and LINDSTRÖM, M. 1957. Arenigian and Llandeilian cherts identified in the Southern Uplands by means of conodonts etc. *Trans. Edinburgh Geol. Soc.*, Vol. 17, 60–70.

LAPWORTH, C. 1889. On the Ballantrae rocks of South Scotland, and their place in the Upland sequence. *Geol. Mag.*, (3), Vol. 6, 20–24, 59–69.

LEGGETT, J. K., McKERROW, W. S. and EALES, M. H. 1979. The Southern Uplands: a Lower Palaeozoic accretionary prism. *J. Geol. Soc. London*, Vol. 136, 755–770.

LINTON, D. L. 1951. Problems of Scottish scenery. *Scott. Geogr. Mag.*, Vol. 67, 65–85.

McADAM, A. D. 1978. Sand and gravel resources of the Lothian Region of Scotland. *Rep. Inst. Geol. Sci.*, No. 78/1.

— and TULLOCH, W. 1985. Geology of the Haddington district. *Mem. Geol. Surv. G.B.*

MACDONALD, R. 1975. Petrochemistry of the early Carboniferous (Dinantian) lavas of Scotland. *Scott. J. Geol.*, Vol. 11, 269–314.

MACGREGOR, A. G. 1928. The classification of Scottish carboniferous olivine-basalts and mugearites. *Trans. Geol. Soc. Glasgow*, Vol. 18, 324–360.

— 1939. The term 'Plagiophyre'. *Bull. Geol. Surv. G.B.*, No. 1, 99–103.

McKERROW, W. S., LEGGETT, J. K. and EALES, M. H. 1977. Imbricate thrust model of the Southern Uplands of Scotland. *Nature, London*, Vol. 267, 237–239.

MARTIN, N. R. 1955. Lower Carboniferous volcanism near North Berwick. *Bull. Geol. Surv. G.B.*, No. 7, 90–100.

MIDGLEY, H. G. 1946. The geology and petrology of the Cockburn Law Intrusion, Berwickshire. *Geol. Mag.*, Vol. 83, 49–66.

MITCHELL, G. H., WALTON, E. K. and GRANT, D. (Editors). 1960. *Edinburgh geology, an excursion guide.* (Edinburgh and London: Oliver and Boyd.)

— and MYKURA, W. 1962. The geology of the neighbourhood of Edinburgh (3rd Edition). *Mem. Geol. Surv. G.B.*

NEVES, R., GUEINN, K. J., CLAYTON, G., IOANNIDES, N. S., NEVILLE, R. S. W. and KRUSZEWSKA, K. 1973. Palynological correlations within the Lower Carboniferous of

Scotland and Northern England. *Trans. R. Soc. Edinburgh*, Vol. 69, 23–70.

— and IOANNIDES, N. 1974. Palynology of the Lower Carboniferous (Dinantian) of the Spilmersford Borehole, East Lothian, Scotland. *Bull. Geol. Surv. G.B.*, No. 45, 73–97.

PEACH, B. N. and HORNE, J. 1899. The Silurian rocks of Britain, Vol. 1. Scotland. *Mem. Geol. Surv. G.B.*

RAGG, J. M. and FUTTY, D. W. 1967. The soils of the country around Haddington and Eyemouth. *Mem. Soil Surv. G.B.*

RICHEY, J. E., ANDERSON, E. M. and MACGREGOR, A. G. 1930. The geology of North Ayrshire. *Mem. Geol. Surv. G.B.*

RICKARDS, R. B. 1976. The sequence of Silurian graptolite zones in the British Isles. *Geol. J.*, Vol. 11, 153–188.

ROCK, N. M. S. and RUNDLE, C. C. *In press.* Evidence for Lower Devonian age of the Great Conglomerate, Scottish Borders. *Scott. J. Geol.*

SISSONS, J. B. 1967. *The evolution of Scotland's scenery.* (Edinburgh: Oliver and Boyd.)

SNODGRASS, C. P. 1953. *The county of East Lothian, The Third Statistical Account of Scotland.* (Edinburgh: Oliver and Boyd.)

SUTHERLAND, D. S. (Editor.) 1982. *Igneous rocks of the British Isles.* (Chichester: Wiley.)

TOMKEIEFF, S. I. 1937. Petrochemistry of the Scottish Carboniferous–Permian igneous rocks. *Bull. Volcan.*, (2), Vol. 1, 59–87.

UPTON, B. J. G., ASPEN, P. and CHAPMAN, N. A. 1983. The upper mantle and deep crust beneath the British Isles: evidence from inclusions in volcanic rocks. *J. Geol. Soc. London*, Vol. 140, 105–121.

WALKDEN, G. M. 1974. Palaeokarstic surfaces in Upper Viséan (Carboniferous) limestones of the Derbyshire Block, England. *J. Sediment. Petrol.*, Vol. 44, 1232–1247.

WALKER, F. 1925. Four granite intrusions in south-eastern Scotland. *Trans. Edinburgh Geol. Soc.*, Vol. 11, 357–365.

— 1928. The plutonic intrusions of the Southern Uplands east of the Nith valley. *Geol. Mag.*, Vol. 65, 153–162.

WALTON, E. K. 1961. Some aspects of the succession and structure in the Lower Palaeozoic rocks of the Southern Uplands of Scotland. *Geol. Rundsch.*, Vol. 50, 63–77.

WHITTINGTON, H. B., DEAN, W. T., FORTEY, R. A., RICKARDS, R. B., RUSHTON, A. W. A. and WRIGHT, A. D. 1984. Definition of the Tremadoc Series and the series of the Ordovician System in Britain. *Geol. Mag.*, Vol. 121, 17–33.

WILSON, H. H. 1952. The Cove Marine Bands in East Lothian and their relation to the Ironstone Shale and Limestone of Redesdale, Northumberland. *Geol. Mag.*, Vol. 89, 305–319.

WILSON, R. B. 1966. A study of the Neilson Shell Bed, a Scottish Lower Carboniferous marine shale. *Bull. Geol. Surv. G.B.*, No. 24, 105–130.

— 1974. A study of the Dinantian marine faunas of south-east Scotland. *Bull. Geol. Surv. G.B.*, No. 46, 35–65.

INDEX

BRITISH GEOLOGICAL SURVEY

Keyworth, Nottinghamshire NG12 5GG

Murchison House, West Mains Road, Edinburgh EH9 3LA

The full range of Survey publications is available through the Sales Desks at Keyworth and Murchison House. Selected items are stocked by the Geological Museum Bookshop, Exhibition Road, London SW7 2DE; all other items may be obtained through the BGS Information Point in the Geological Museum. All the books are listed in HMSO's Sectional List 45. Maps are listed in the BGS Map Catalogue and Ordnance Survey's Trade Catalogue. They can be bought from Ordnance Survey Agents as well as from BGS.

On 1 January 1984 the Institute of Geological Sciences was renamed the British Geological Survey. It continues to carry out the geological survey of Great Britain and Northern Ireland (the latter as an agency service for the government of Northern Ireland), and of the surrounding continental shelf, as well as its basic research projects. It also undertakes programmes of British technical aid in geology in developing countries as arranged by the Overseas Development Administration.

The British Geological Survey is a component body of the Natural Environment Research Council.